A catchment-based approach to
Soil and Water Conservation

A focus on the Siwalik Hills of Nepal Himalaya

Sohan Ghimire PhD

The James Hutton Institute, Aberdeen, Scotland, UK

Daisuke Higaki PhD

Faculty of Agriculture and Life Science, Hirosaki University, Japan

First published 2012

Copyright © 2012 Sohan Ghimire and Daisuke Higaki

All rights reserved. No part of this publication may be reproduced or transmitted in any form or by any means, electronically or mechanically, including photocopying, recording or any information storage or retrieval system, without either prior permission in writing from the publisher, or the author except in the case of brief excerpts used in critical articles and reviews.

Unauthorised reproduction of any part of this work is illegal and is punishable by law.

The advice and information in this book are believed to be true and accurate at the date of going to the press, but neither the author nor the publisher can accept any legal responsibility or liability for any errors or omissions.

ISBN: 978 – 1-4717-4353-5

Published by Siwalik Hills Research & Development Group (SIRED)

G.P.O.Box 9200, Kathmandu, Nepal

www.siwalikhills.org

Printed by lulu publishing Ltd. (www.lulu.com).

Contents

	Preface	vii
	Acknowledgements	viii
1	**Introduction**	1
	Background	1
	Key environmental issues	5
	Key processes relating to soil and water mobilisation	7
	Objectives of this book	9
2	**Geomorphology of Siwalik Hills**	11
	The Siwalik Hills	11
	Trijuga River valley	16
	Stream systems	17
	Experimental catchment of Khajuri stream	21
	Geomorphological classification	21
	Drainage patterns	24
3	**Research design, Field measurements and techniques**	25
	Research design	25
	Field measurement	25
	Techniques	28
	Study of aerial photographs	28
	Erosion pin method	29
4	**Changes in land cover and stream planform**	31
	Background	31
	Land cover change in Trijuga River valley	32
	Temporal and spatial variations	32
	Land cover change in Khajuri catchment	35
	Analysis of stream planform change	36
	Changes in planform in the experimental streams	39

		Effect of land cover change in stream planform change	40
		Stream planform change processes	41
		Driving factors of stream pattern change	44
		Implications of land cover and stream form changes	45
		Summary	46
5	**Rainfall characteristics**		48
		Rainfall trends in Nepal	48
		Annual and seasonal rainfall patterns	49
		Daily rainfall patterns	51
		Major rainfall events	53
		Frequency analysis and return periods	53
		Summary	55
6	**Hydrology, flood and sediment**		56
		Background	56
		Measurement of flood discharge	56
		Rainfall-runoff relationship	58
		Infiltration characteristics	61
		Estimation of peak discharge	63
		Measurement of suspended sediment	65
		Summary	67
7	**Identification of sediment sources**		68
		Key sources of sediment	68
		Distribution of sediment sources	69
		Summary	71
8	**Hillslope erosion processes**		72
		Surface erosion	72
		Layout and design of erosion plots	72
		Rate and process of erosion	74
		Controlling factors	77

Summary	78
Rill erosion	79
Forms and processes	79
Measurement of rill erosion	80
Summary	82
Gully erosion	83
How are gullies formed?	83
Types and processes of gully erosion	84
Previous studies	85
Characteristics of study gullies	89
Gully head enlargement	92
Measurement of gully head erosion	97
Estimation of sediment production	100
Summary	103
Landslide slope erosion	104
Processes of erosion	105
Mapping and field measurement of landslides	107
Summary	108
9 Stream bank erosion	**109**
Background	109
Measurement techniques	110
Process of bank erosion	111
Vegetation and bank erosion process	112
Issues of bank erosion	113
Streambank characteristics	115
Measurement of streambank erosion	118
Temporal and spatial variations	118
Controlling factors of bank erosion	119
Streambank failure mechanism	122
Streambank-vegetation interaction	124
Channel course change through cutoff	125

	Bank erosion hazard mapping	126
	Summary	132
10	**Estimation of soil loss at a catchment scale**	134
	Surface erosion	134
	Gully erosion	135
	Landslide slope erosion	135
	Streambank erosion	137
	Catchment-wide sediment production	139
	Erosion hazard map	140
	Sediment yield	142
	Assessing degradation status of the catchment	144
	Summary	145
11	**Conservation efforts and management policy**	146
	Management policy	146
	Some examples of conservation schemes	147
	Key issues about soil and water conservation	151
	Summary	154
12	**Lessons learned**	155
	Synthesis of findings	155
	Lessons learned	155
	Recommendations	167
	References	169

Preface

Field-based research works hold a significant importance especially in the field of soil and water conservation management. However, there is a limited availability of field-based monitoring data and information on the conservation of the Siwalik Hills of Nepal Himalaya as researchers have overwhelmingly concentrated their efforts on understanding of ecology/environment of the High Himalaya and Mountain regions. In view of the importance of the Siwalik Hills, being a source of water and sediment for lowland plain of Terai, there is a high need to provide focus on the field-based research works in this region. This study is a kind of this. An integrated approach has been adopted to identify the key issues related to water and sediment, understand the operating processes and key drivers of change and quantify them where possible in order to develop tools and methods that can be useful for assessing the status of land degradation and water/sediment hazard condition. So, on one hand, this study focuses on a unit-level study component such as a process of gully head retreat while on the other hand it derives many policy conclusions based on the collective analysis of several components. The study does not present the solutions to all the issues, but it develops a number of techniques and methods to help address some of the key issues identified by the study. We hope that the approaches and methods proposed in this book will provide a basis for the overall conservation of soil and water in the Siwalik Hills.

Sohan Ghimire
Daisuke Higaki

11 June 2012

Acknowledgements

We are grateful to the Ministry of Education, Sports and Culture of Japan for providing a fellowship to the first author for pursuing PhD which forms the key part of this book. The Sabo Technical Centre (STC) of Japan provided financial support for the field works. We kindly acknowledge this support.

We would like to thank Prof. Vishnu Dangol, Professor of Department of Geology, Tribhuwan University Nepal and his student Mr Tara Bhattarai for the support and co-operation throughout the study. We also thank to the colleagues of the Department of Water-Induced Disaster and Prevention (DWIDP) for their cooperation during Nepal visits for the field works.

The help and support provided by the James Hutton Institute, Aberdeen, Scotland during the preparation of the manuscript is gratefully acknowledged.

We dedicate this book to the Mushahar people who belong to one of the most disadvantaged and marginalised ethnic groups of Nepal living on the foothills of the Siwalik Hills.

1
Knowing soil and water issues in Siwalik Hills

Background

Soil erosion and consequent land degradation problems by rainfall waters are widely documented especially through the steep slope lands and agricultural terraces in Nepal Himalaya. Deforestation, overgrazing and intensive agriculture are often considered the principle causes of land degradation and wider environmental crisis in the region.

Human activities are often attributed as the cause of the increased land degradation and environmental crisis (Eckholm, 1976), however, some studies such as by Ives and Messerli, 1989 outlined the main causes to the natural processes. They argue that whereas serious environmental and human problems exist in the Himalayan region, their impact on the downstream areas cannot be proven. In general, the impact of human intervention is a question of scale (Ives and Messerli, 1989; Lauterburg, 1993). The effects of human intervention in the large catchments are generally not visible (FAO, 2002), while it is likely to be noticeable at the smaller spatial scales. However, studies to document the extent of the problems on these aspects are lacking.

While the water and sediment related issues are widely studied mainly in the Middle Mountain region (Mountain range in between the Siwalik Hills and Lesser Himalaya) of Nepal, very few studies have been undertaken in the Siwalik Hills even though the environmental issues in the latter are not less serious. The Siwalik Hills are situated in between the Indo-Gangtic plain to the south and Middle Mountain to the north, which are known as the foothills of the Himalayas and a zone of active crustal movement (Valdyia, 1998; Kizaki, 1988). Compounded with intense rainfall and steeper topography, the area is believed to be developed with the evolution of unstable slopes comprising vigorous rills, gullies and stream channels. Proportion of these unstable slopes in the Siwalik Hills was found ten times higher than in the Middle Mountain region (LRMP, 1986), even though both regions belong to the similar monsoon rainfall regime.

Because of the active erosion processes occurring on the unstable slopes, a huge amount of sediment is being lost from the catchments (Upreti, 2001) leading to land degradation in the region as well as sediment deposition in the plains of Terai. The Terai zone is the main backbone of the economy of the country because it supplies more than two-third of the food grains in the country. Hence, protection of the Siwalik Hills is also meant the protection of the Terai plains.

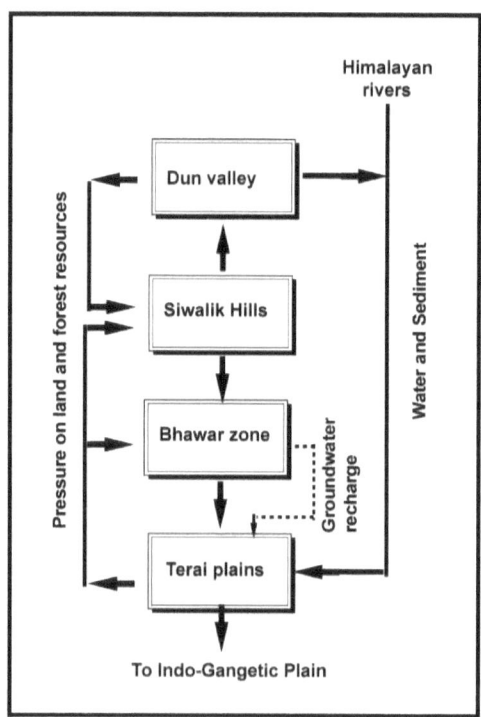

Figure 1.1 Conceptual diagram showing inter-linkage between Siwalik Hills and Terai plain in terms of land and water use.

There must be a high level of dependency and hence direct linkage between the Siwalik Hills and Terai plains regarding water and sediment transport system because of unique system of river flow and proximity in the location (Fig. 1.1).

The Siwalik-Terai inter-linkage model can be explained by considering the two more components called Dun valley and Bhawar zone. Dun valley is the lowland alluvium formed where the Siwalik Hills bifurcate. The foothill of the Siwaliks, which is called Bhabar zone, is mostly composed of coarse boulders. This is considered to be the important recharge area, which is the main source of underground water in the Terai plains (Lauterburg, 1993). All the river systems in the Himalaya ultimately cross the lowland plains of Terai to join the Bay of Bangal through the Indo-Gangetic plain. Two types of stream systems occur in the Siwalik Hills. One type streams discharge into the large scale perennial rivers before flowing onto the Terai plains. Other categories of streams directly flow towards the Terai plain via Bhawar zone.

In many locations, Bhawar zone consists of forest (popularly known as "Char Kose Jhadi" in the eastern Nepal) and shrublands. Thus, the forest resources of the Bhawar and Siwalik Hills are under the increasing pressure from the increasing population in the Dun valley and Terai plain. However, virtually no study has been undertaken so far regarding this kind of inter-dependent biophysical systems. These facts suggest that it is urgent to gain understanding of the environmental issues in the Siwalik Hill. Further reasons why the issues relating to Siwalik Hill are important have been summarised in Box 1.1.

Some studies in the past have pointed out environmental degradation in the Siwalik Hills mainly through soil erosion. Most of them were concerned with only surface erosion estimation, often with short term monitoring (eg. Laban, 1978) and information on long-term soil loss is scanty. A study done by Honda et al. (1996) evaluated the soil yield from a Siwalik catchment called "Ratu" using the technique of GIS/RS. They found denudation rate of more than 3 mm per year which is reasonably higher than the average rate of soil loss from Nepal (1.63 mm) (DSCWM, 2004). Subsequently, Samarakoon et al. (1996) pointed out considerable sedimentation and deviation in stream channel in the floodplain of the same river using sequential satellite images, attributing the main cause to the massive sediment production from the degraded catchment.

> **Box 1.1 Reasons why a focus on understanding the geo-environmental issues in the Siwalik Hills is needed.**
>
> - Environmental degradation in the Siwalik Hills is believed to be the result of weak geological formation coupled with the effects from human activities. However, field measured quantitative data to support this hypothesis is lacking.
> - Water and sediment problems in terms of flood disasters are very common in this region. However, evidence-based studies on the mechanisms of flash flood and flow hydraulics are very rare.
> - Research studies on the estimation of the sediment production from the catchment are also lacking. Without knowledge on the extent of sediment production, storage and conveyance, no effective mitigation works can be proposed.
> - The Siwalik Hill is shorter and narrower ecological zone compared to other ecological zones in the Himalaya. Due to this reason, the catchments in this region are much smaller in size and the streams are shorter in length. More knowledge on the process interactions can be obtained from the study undertaken in smaller scales than in larger spatial scales.
> - Generally all types of soil erosion occur in the Siwalik Hills. This provides an opportunity to undertake study on various sediment processes within a single catchment, which substantially reduces time, manpower, equipment and management.
> - There is a better opportunity for the application of findings in other areas because the region is more homogeneous in terms of geomorphology and land use compared to other regions in the Himalaya.

There is a lack of study which focuses on mass wasting processes in sediment yield estimation from Siwalik Hills catchments, even though soil loss by mass movement is the most significant type of erosion in the steep hill catchments (Silby, 1993; Knighton, 1998). Also, these processes have not been documented comprehensively.

A detailed understanding on the condition and processes related to sediment mobilisation and stream behaviour is still lacking. In this context, this study has attempted to generate and use field-based data collected over a three year period with intensive monitoring on the main aspects of soil and water mobilisation from the Siwalik Hills. The monitoring data are complemented by undertaking a study using remote sensing data and aerial photographs.

Key environmental issues

A questionnaire survey was conducted in a study area (Khajuri Khola in Udayapur district, Chapter 2) in June 2003 relating to the key environmental issues in the area. Of randomly selected 30 respondents, almost all identified the flood issue as their major concern. Frequent damage to the cropland and constant threat of damage to their houses were the reasons why flood was their primary concern. About 80% of the respondents outlined the water scarcity problem especially during the winter period. Most of them were quite aware of the erosion problems in the upland catchments, especially in the large gullies. About 70 percent expressed their satisfaction over the achievements of community forestry that started from a decade ago, even though perfect control on grazing and fodder cutting was still not possible. Only a few of the respondents were well aware of the water quality and pollution problem in the stream especially from industrial activities in the area.

From the study, four major issues were identified important and their order of importance is presented in Table 1.1. It is important to note that even though the priority order of the issues may be site-specific, the issues outlined in the table may not exist in every area in the Siwalik Hill.

Table 1.1 Key issues in the Siwalik Hills (based on survey in 2003 in the study area).

Issue	Priority
Flood	1
Water scarcity	2
Erosion problems	3
Forest management problem	4

The most important issue is related to stream floods. The floods are often of erratic nature and high potential that may cause damage directly to the people or adjacent lands (Fig. 1.2). During the rainy season, the flooded streams do not only interrupt the public movement, but also sometimes halt the traffic movement bringing more effects to the socio-economic conditions.

The second issue- water scarcity is associated with the scarce water resources available in the Siwalik Hills. Water scarcity is found both for drinking and irrigation purposes, mostly during the dry season. Since the streams are of non- perennial types that run completely dry during the time of no rainfall, they cannot supply enough water required for domestic purposes.

The third issue related to soil erosion is mainly concerned with the degradation of headwater catchment through various soil erosion and mass wasting processes. The Siwalik Hills zone has been plagued by complicated institutional, natural and anthropogenic problems, the most serious of which are soil erosion, degradation of catchment areas, and diminished productivity (Oli, 1999).

The land degradation issue is often linked with the land cover. Since the headwater of the Siwalik Hills mostly consists of vegetation cover (generally called rain forest), the issues regarding the sustainable use of forest resource obviously stand as important. Land cover change, particularly conversion of forest into agricultural land (i.e. deforestation) is the main issue at the moment, which is often linked as the main cause of the environmental degradation in the Siwalik Hills.

Figure 1.2 A village situated on the foot of the Siwalik Hills in Sindhuli district was badly destroyed by a flash flood in July 1995 (Photo: DWIDP, Kathmandu).

Like in other parts of Nepal Himalaya, massive deforestation occurred in the recent past in the Siwalik Hills (MOFSC, 1994). Schweik et al. (1997) found severe deforestation in two small basins in the Siwalik Hills from 1978 to 1992. Honda et al. (1996) also pointed out the problem of deforestation in the catchment of the Ratu river.

Over the past few decades, the Government of Nepal has provided a priority on the preservation of the environment of Siwalik Hills, in particular the forest cover. Like in other parts of the country, management policy relating to community forestry has been started as per the Forest Act of 1993, which focuses on the handing over forest to local user groups. Even though some researches pointed out the positive outcomes of this policy in some areas (e.g., Gautam et al., 2002), there is no scientific assessment available on its effectiveness on lessening the land degradation problems.

Key processes relating to soil and water mobilisation

Higaki (2003) outlines the history of geomorphological development, evolution of river catchment and erosion processes in the Siwalik Hills (Fig. 1.3). Since mudstone or sandstone of the Lower and Middle Siwaliks and conglomerate of the Upper Siwaliks of fluvial origin are unconsolidated (Sharma, 1990), weathering and various hillslope erosion processes actively proceed. Geological layers of the Siwaliks generally dip towards north by the crustal movement since the Middle Pleistocene (Sharma, 1990; Kimura, 2000). Evolution of river system started since then. The terrain formed as alluvial fans of these rivers had been up-lifted, thus forming fluvial terraces at present at the middle reach of the rivers. Downstream area had been transformed into alluvial plain due to crustal subsidence. Consequently present geomorphological setting of the Siwalik Hills can be divided into three segments: (a) Hills with densely distributed drainage, (b) Fluvial terrace area, and (c) Alluvial plain. Segment (c) can be divided into flood plain and alluvial fan along the stream.

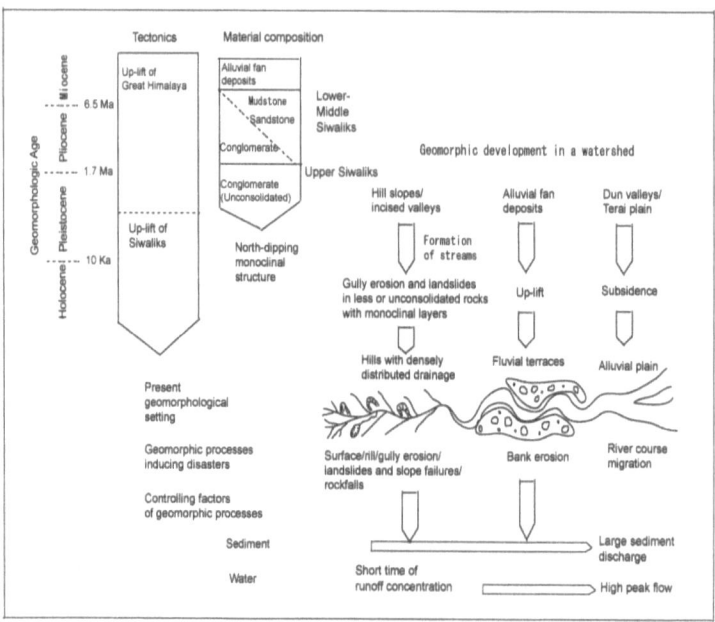

Figure 1.3 Geomorphological development, river channel evolution and erosion processes in Siwalik Hills (Higaki, 2003).

Highly dense drainage system with steep incised valleys was formed by gully erosion, slope failures and landslides in unconsolidated and weathered layers in segment (a). This results in short time of run-off concentration downstream. Landslides often occur on the north-facing dip-slopes in this segment. Since boulders and gravels that form the fluvial terrace banks and riverbeds originated form conglomerates of the Upper Siwaliks, particle size of bank materials is not large. This results in active bank erosion in segment (b) (Higaki, 1998). These can be the major sources of high peak flow and sediment discharge downstream. Hence inundation and river course migration tend to occur in segment (c).

In this way, it is clear that the catchments in the Siwalik Hills have undergone different sets of actions and processes that result in water and sediment mobilization in different ways. Consequently, extent and type of potential effects also vary widely in relation to the catchment (or stream) segment as shown in Fig.1.3. Therefore, a catchment classification based on geomorphological characteristics can be a useful tool for understanding the land degradation and sediment mobilisation issues in the Siwalik Hills.

Objectives of this book

This book focuses on understanding of an integrated approach which has been applied to identify and understand the soil and water conservation issues through various water erosion processes in the micro-scale catchments of the Siwalik Hills. The analysis and discussion will mainly focus on the hillslope and fluvial environments in the catchment. The book presents findings from a rigorous field-based study that aimed to investigate the extent, distribution, rates and processes of sediment generation with some deliberations at the causes and potential effects within the catchment.

The main objective of this book is to identify main sources of sediment in the hillslope and fluvial systems, and to explain about key drivers, processes and impacts within a catchment. Various sediment generating processes have been analysed and discussed in this book which include surface, rill, gully and landslide slope erosion, along with from stream channels with a focus on streambank erosion.

The field-based study which forms the key part of this book had the following specific objectives:

(1) To examine the long-term stream pattern changes as a result of erosion processes and to find out if there was any link with land cover change.
(2) To investigate flash-flood mechanism and its controlling factors.
(3) To identify type and distribution of sediment sources within the catchment
(4) To determine the annual average rates of sediment production within the catchment
(5) To find out the annual average rates, operating processes and controlling factors of streambank erosion.
(6) To develop a methodology framework for the assessment of bank erosion hazard based on field observation.
(7) To assess the sediment contribution from each type of source and estimate the annual total sediment production rate from the catchment.
(8) To develop a map of erosion sources based on field measurement.
(9) To synthesise and analyze the measured data, information and relevant processes in order to assess their implications for wider management policies for addressing the soil and water conservation issues in the Siwalik Hill region.

The framework and the study objectives are presented in Fig. 1.4. The study broadly comprises two spatial scales. First, we investigate the general trend and behaviour of the stream systems in a regional scale by undertaking historical change analysis on land cover and stream pattern. The main purpose of this is to gain better understanding of the nature and type of changes that can be useful for understanding present status and relevant processes. The study then investigates the processes operating currently in a selected catchment to understand sediment generation processes in a more detailed way.

Each type of sediment source and operating process has been analysed based on ground monitoring data. The analysis and discussions are centered on the geomorphological classification of the catchment. The ultimate aim is to evaluate the degradation status of the catchment based on the extent and nature of soil and water mobilisation from the catchment.

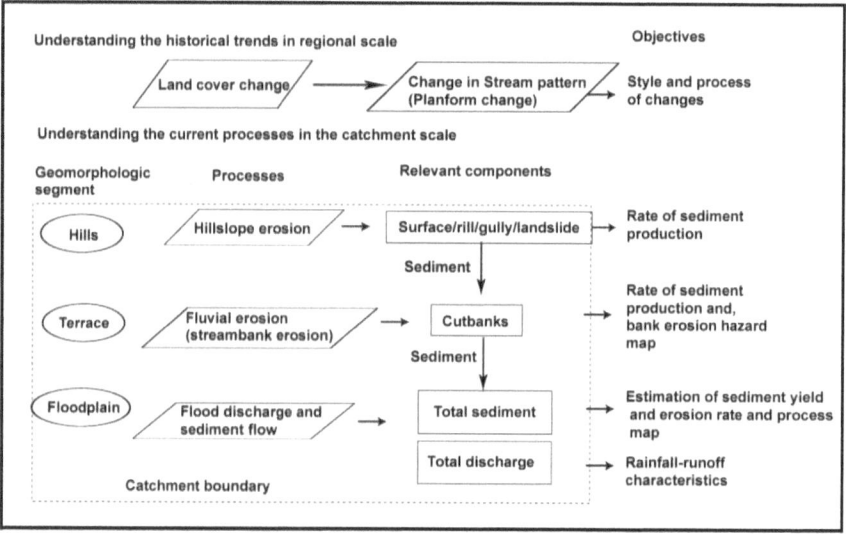

Figure 1.4 Framework of the study.

On the basis of the framework above, this book tries to explore the following questions:
(i) What are the main issues and processes that are responsible for the land degradation?
(ii) How much is the contribution of each process to total sediment yield?
(iii) What are the governing parameters of the erosion processes?
(iv) What is the effect of human activities on the status/processes in the catchment, and particularly in the stream systems?
(v) What is the status of the catchment in terms of land degradation?
(vi) What types of management policies are appropriate for addressing the environmental issues in the Siwalik Hills?

2
Geomorphology of Siwalik Hills

The river catchment provides a clearly defined physical unit for a wide variety of studies. It contains the recognizable elements of an open system with inputs, throughputs and outputs of energy and matter. The character and behaviour of the fluvial system at any particular location reflect the integrated effect of a set of upstream controls, notably climate, geology, land use and basin physiography, which together determine the hydrologic regime and the quantity and type of sediment supplied (Knighton, 1998).

Dealing with these environment controls, spatial context of the study area is understood. There are broadly three level of scales described in this chapter: general introduction of the Siwalik Hills, regional characteristics particularly stream systems in the Trijuga river valley and the experimental catchment of Khajuri stream.

The Siwalik Hills

The Siwalik Hills are located within the political boundaries of Pakistan, India, Nepal, and Bhutan (Acharya, 1994) (Fig. 2.1). They gradually become steeper and narrower in relief and width respectively, from northern Pakistan to Bhutan (over 2000 km in length).

Nepal can be divided into eight well-defined physiographic units running roughly east west (Fig. 2.2, Table 2.1). The Siwalik Hills belong to the physiographic unit that lies between the Terai Plain in the south and Mahabharat Range in the north. Generally they rise abruptly from the plains of the Terai. Running the length of Nepal, the hills typically range in elevation from 200 to 1000 m, and rise to 1300m in many places. They have an arc type face with conspicuously north-dipping beds, forming a steep escarpment towards the Terai (Upreti, 2001).

Geomorphologically, the Siwalik Hills exhibit a very immature topography with highly rugged terrain dissected by numerous gullies (LRMP, 1986; Upreti, 2001). Ongoing erosion and tectonic activity has greatly affected the topography of the Siwaliks. Their present-day morphology is comprised of hogback ridges and gullies with other many kinds of erosional features (Mukerji, 1976). Associated badlands features include the lack of vegetation, steep slopes, high drainage density, and rapid erosion rates (Howard, 1994).

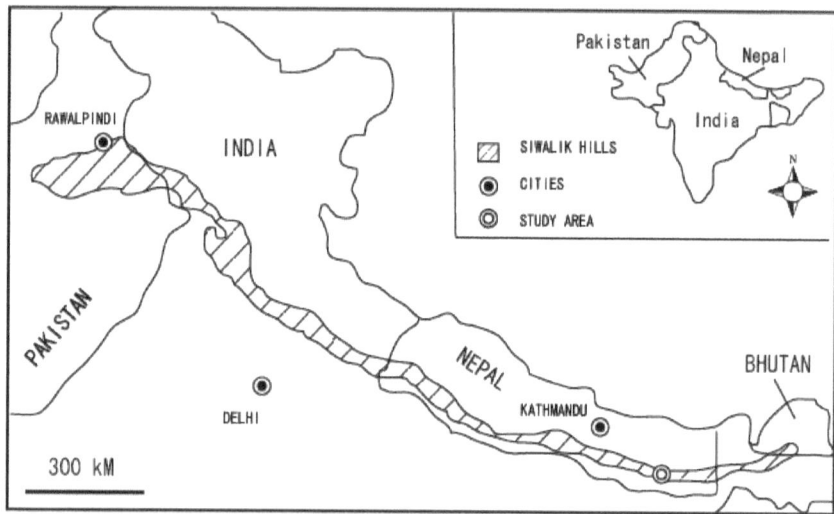

Figure 2.1 The Siwalik Hills extending east-west along Nepal and India.

The Siwalik zone consists of fluvial sedimentary rocks of Neogene to Quaternary age (14 to 1 million years ago). The zone is bounded to the north by the *Main Boundary Thrust (MBT)* and to the south by the *Main Frontal Thrust (MFT)*. In general, the rocks of the Siwalik zone dip northwards at varying angles and the overall strike is east-west.

The rocks of this zone are divided stratigraphically into three parts: Lower, Middle and Upper Siwaliks (Upreti, 2001). The deepest stratum, which is called the Lower Siwalik, is essentially composed of alternations of fine-grained sediments such as various coloured mudstone, siltstone and shale with subordinate amounts of fine-grained sandstone. The Middle Siwalik is marked by the first appearance of thick multistoried sandstone beds and alternating with subordinate beds of mudstone. The Upper Siwalik is characterized by very coarse-grained rocks such as boulder conglomerates with minor proportions of mudstone intercalation.

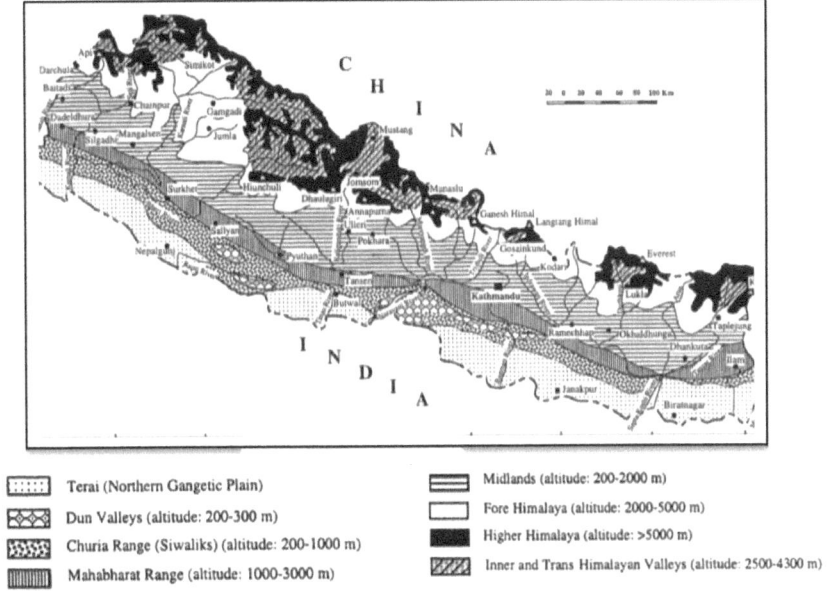

Figure 2.2 Physiographic subdivisions of Nepal (Upreti and Le Port, 1999).

The parent material of the soil is the uplifted stratified alluvium. The textures of the soil are directly related to the underlying bedrock geology with medium textured soils on mudstone, siltstones and coarse textured soils with boulders on conglomerates. Weak geology and steep topography are responsible for severe erosion despite good forest cover.

The extent of agricultural land is very limited. Without irrigation, droughty soils give low yields. Constant burning and grazing are having a serious effect on soil quality (LRMP, 1986). Main features of the Siwalik Hills pointed out by (LRMP, 1986) are given in Box 2.1

Table 2.1 Physiographic units of Nepal (Upreti, 2001).

Physiographic unit	Width (km)	Altitude range (m)	Main rock types	Climate
Terai Plain	10-50	100-200	Recent Alluvium	Sub-tropical
Siwalik Range	10-50	200-1300	Sandstone, Mudstone, Shale, Conglomerate	Sub-tropical
Dun Valley	5-30	200-300	Coarse to fine alluvium sediment	Sub-tropical
Mahabharat Range	10-35	1000-3000	Schist, Phyllite, Gneiss, Quartzite, Limestone	Temperate
Midlands	40-60	200-2000	Schist, Phyllite, Gneiss, Quartzite, Limestone	Sub-tropical to Temperate
Fore Himalaya	20-150	2000-5000	Schist, Phyllite, Gneiss, Marble	Sub-alpine
Higher Himalaya	10-60	>5000	Schist, Gneiss, Migmatites, Marbles	Alpine
Inner and Trans Himalayan Valleys		2500-4300	Schist, Gneiss, Migmatites, Marbles and Tethyan sediments	Alpine

Dun valley

Intermittently located between the Siwaliks and the Lesser Himalayas or Mahabharat Hills (exclusively in India and Nepal) are Dun with flat-bottomed longitudinal structural valleys with their own drainage systems (Nakata, 1972). These essentially comprise several large Himalayan piedmont alluvial fans and terraces, which formed as a result of tectonic episodes in the flanking Siwaliks. In Nepal, these duns were often naturally filled with alluvial sediments of lacustrine and fluvial deposits of Quaternary to Recent (1,8 million years to present) age (Tokuoka et al., 1986).

The altitude of the Dun valleys varies from 200 to 300m. The valley soil being very fertile is extensively used for cultivation. Whereas the Siwalik Hills remain sparsely populated, the population of the Dun valleys has increased markedly in the last few decades (Upreti, 2001). In effect, pressure on the natural resources, particularly agricultural land and forest is persistently increasing.

> **Box 2.1** Key features of the Siwalik Hills (LRMP, 1986).
>
> - Steep, broken terrain.
> - Shallow droughty soils with low surface infiltration and percolation rates.
> - High intensity rainfall during monsoon, and tremendous overflow torrents regardless of vegetation cover.
> - High insolation and air temperatures during the dry season.
> - Lack of water for domestic purposes and irrigation.

Trijuga River valley

The study area called Trijuga river valley is located in the Siwalik region of eastern Nepal (Fig. 2.2). The catchment area of the valley is about 640 sq. km extending east-west with an average length of about 45 km. The valley floor elevation ranges from 130 to 200 m above sea level. It is surrounded by the Siwalik Hills with hogbag features (kimura, 1997).

Geological map of the Trijuga valley is shown in Fig. 2.3. The valley includes three types of formation: lower, middle and upper Siwaliks. The northern part of the hills is mostly composed of fine-grained claystone and siltstone of lower Siwalik formation, while that in the southern part is mostly made up of coarse sandstone and boulder conglomerates of middle and upper Siwaliks (Kimura, 1997). The region is in the state of active crustal movement with occurrence of many faults (e.g. Tokuoka et al., 1986, Nakata, 1982, Schelling, 1992)

Geomorphological assessment and research around the Trijuga valley was carried out by many researchers in the past (e.g. Itihara et al., 1972; Kimura, 1997).These researches mostly focused on geomorphological classification. Kimura (1997) identified four types of terraces based on lithological composition of terrace deposits (Fig. 2.4).

Three types of the terraces occur on the outer Siwalik Hills namely: Bokse 1, Bokse 2 and Bokse 3, while the fourth type- Gaighat terrace mostly occurs in the inner Siwalik Hills.

Due to monsoonal regime, the Siwalik Hills is affected by an extremely humid and hot climate that can be considered to be uniform at the scale of investigation. The annual average rainfall is about 1900 mm.

Whilst most part of catchment's headwaters is covered by natural vegetation i.e. forest and shrubs, areas in the downstream floodplains are used for cultivation. Sal (Shorea Robusta) is the dominant type of tree species in the region. Tropical deciduous riverine forest comprising Khair (Acacia catechu) and Sisso (Dalbergia sissoo) is predominant along the streams (Chaudhary, 1998).

The valley includes a significant area of Udayapur district, where the population is more than 0.2 million (CBS, 2001). A number of ethnic groups have been living in the valley for a long. The most disadvantaged and backward ethnic group called " Mushahar" reside mostly on the banks of the streams.

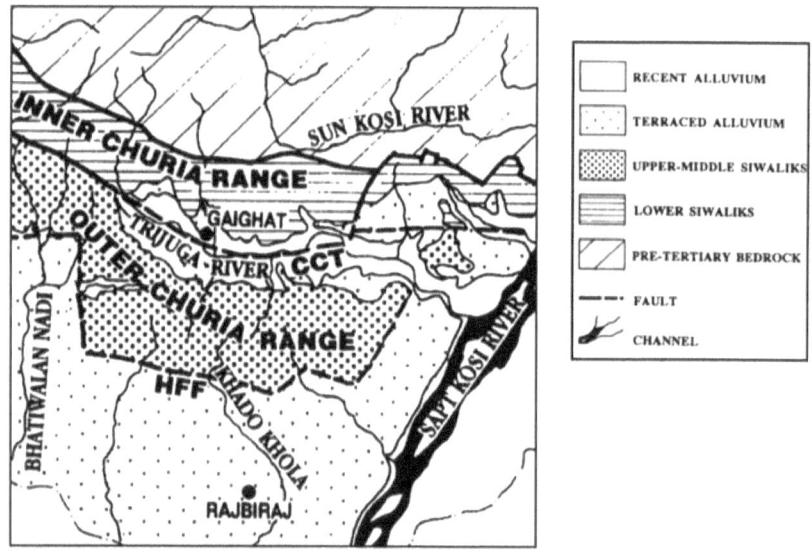

Figure 2.3 Geological map of Trijuga river valley (Kimura, 1997).

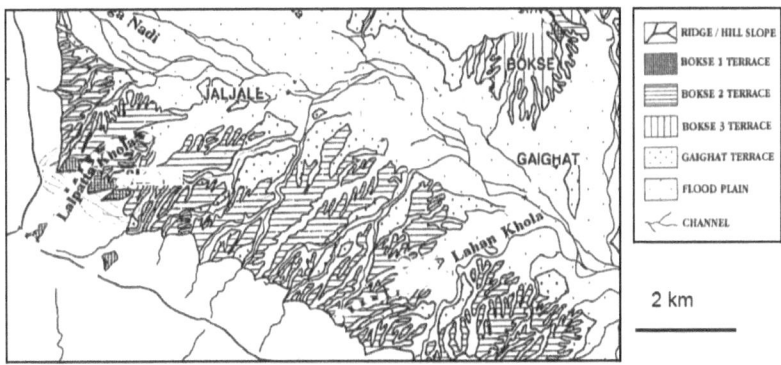

Figure 2.4 Geomorphological classification of the southwest part of Trijuga River valley (Kimura, 1997).

Stream systems

The main river -Trijuga originates from a number of drainage channels dissected in the hillslope of Siwalik Hills before joining the Koshi river to the east (Fig. 2.5). It is important to note that the stream systems, which form the significant proportion of the drainage pattern of the Siwalik Hills, are the main focus rather than the main river. These streams are principally part of headwater drainage channels which are relatively small but steep in slope gradient. Degradation of the headwaters is generally resulted from the development of such channels through various erosion processes. Some selected streams are shown in Fig. 2.5 and their main characteristics are presented in Table 2.2.

The streams drainage area varies from a few square kilometers to as much as 50 square kilometers. Of them, except Baruwa stream which begins from the inner Siwalik Hills, all the streams originate from the outer Siwaliks. Length of the main channel varies from 2 km to 14 km and slope gradient from 1.6 to 3 percent. The drainage density ranges from 3 to 6 km/km^2.

As the catchment area is the dominant control of stream discharge, stream characteristics are analysed taking this factor into consideration (Fig. 2.6). The length of the main channel generally varies linearly with the drainage area. This means that the width/length ratio of the catchments is generally low implying that the shape of the catchment is relatively elongated. The drainage density generally does not show significant variation against the size of the catchment. Average channel gradient also does not show any clear relationship with the drainage area.

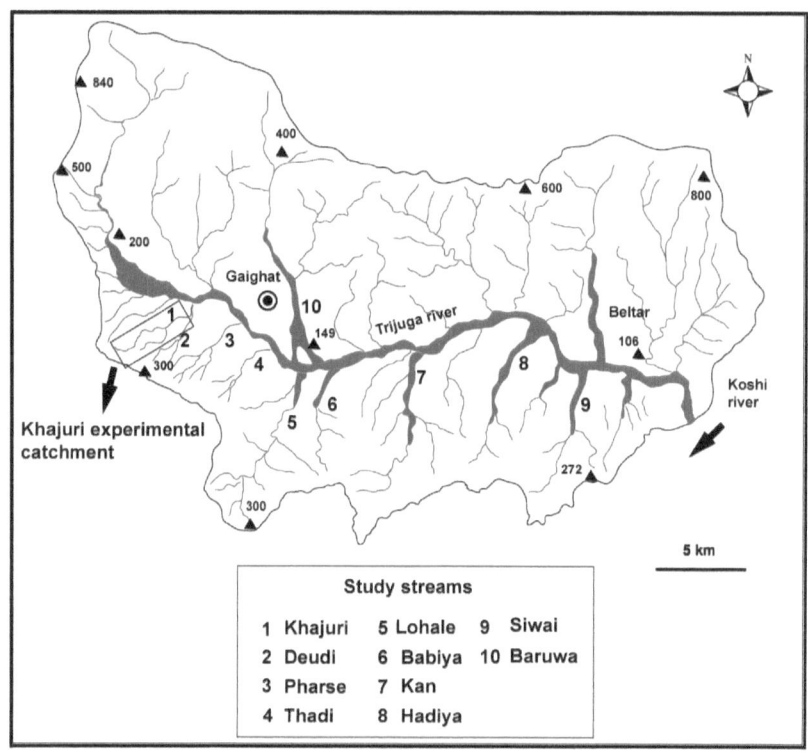

Figure 2.5 Drainage network and study streams in the Trijuga River valley. The spot heights are in metres from mean sea level.

Further characteristics of the stream channels are presented in the subsequent chapters based on field observation and measurement.

Table 2.2 Characteristics of some selected Siwalik streams.

Stream	Drainage area, km^2	Channel L. km	Relative Relief, m	Gradient %	Drainage density km/km^2
Khajuri	2.8	3.68	135	2.0	3.5
Masahar	1.8	2.63	135	2.5	5.5
Deuri 1	1.6	2.50	205	2.6	5.9
Deuri 2	1.3	2.25	200	2.9	4.5
Bagaha	1.5	3.25	195	2.5	5.4
Khahare 1	1.4	3.00	185	2.8	4.1
Khahare 2	2.2	2.75	170	3.1	3.0
Siwai	21.1	6.41	159	2.5	3.5
Hadiya	17.1	8.12	151	1.9	4.0
Kan	33.6	7.97	129	1.6	3.7
Babiya	14.8	6.61	123	1.9	5.2
Lohale	23.0	9.08	154	1.7	4.2
Baruwa	49.0	14.06	401	2.9	5.1

Box 2.2 Local names of ephemeral streams in the Siwalik Hills.

- Khahare Khola (meaning vigorous and turbulent)
- Thado/Thadi Khola (meaning very steep slope)
- Sankaha Khola (meaning very erratic stream)
- Khads (Western Nepal and India)
- Chaoes (in some parts of India)

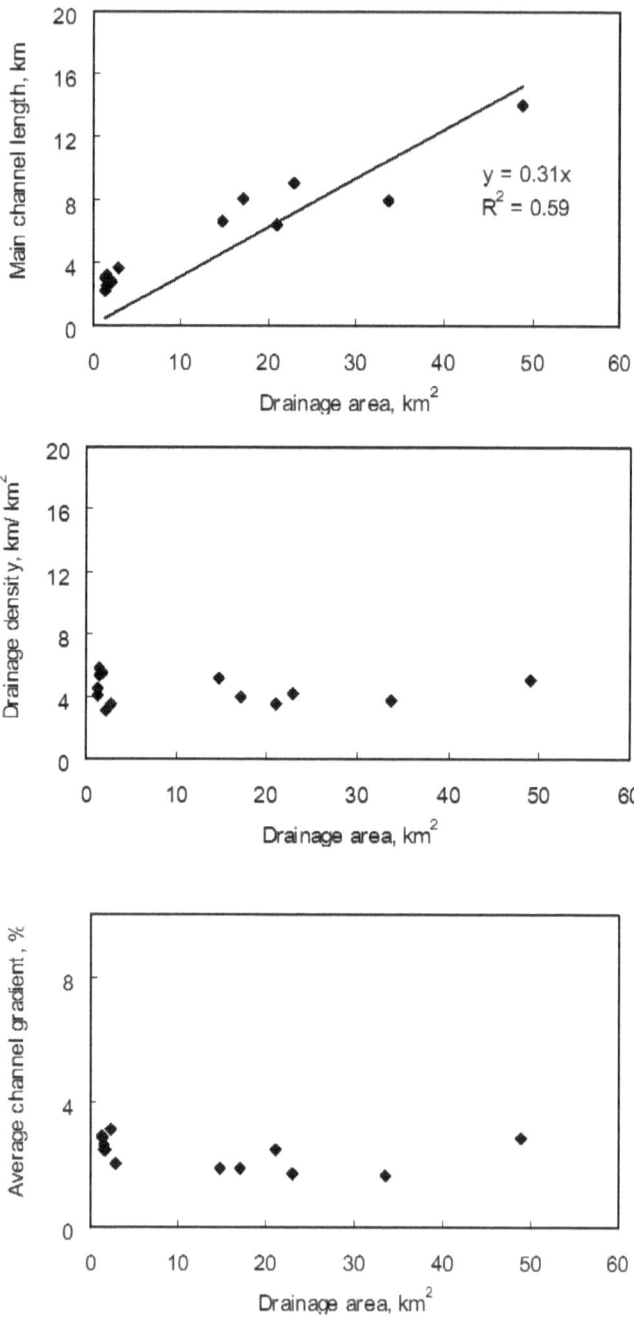

Figure 2.6 Channel properties verses drainage area.

Experimental catchment of Khajuri stream

The experimental catchment selected was the Khajuri Khola (stream), which is located in the south-west of the Trijuga river basin (Fig. 2.5). Elevation from mean sea level varies from 165m on the floodplain to 370m on hilltop. Topography of the catchment and its surrounding is shown in Fig. 2.7, while that of the Khajuri catchment in Fig. 2.8.

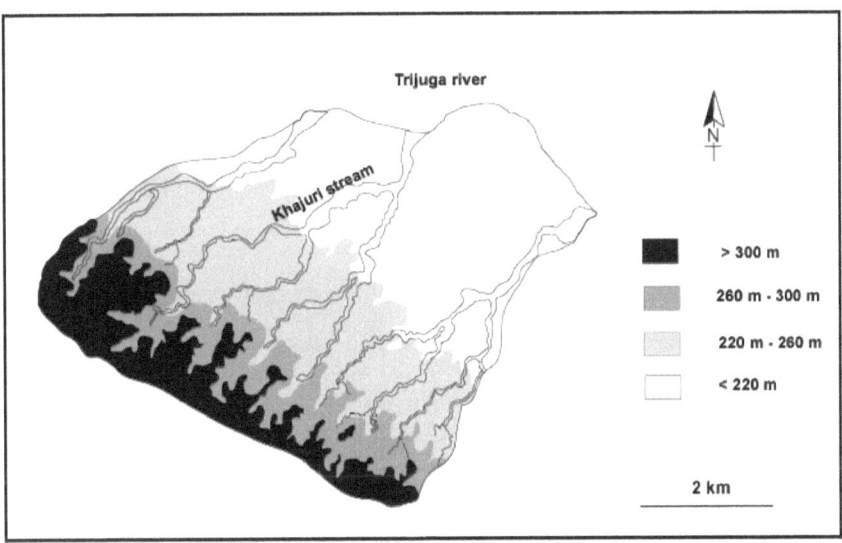

Figure 2.7 Topography of Khajuri catchment and its surrounding area.

The study area covers an area of 4.62 sq km. It consists two small catchments and stream systems (Khajuri and Mushahar) which drain into the Trijuga River. The main hill consists of many sub-ridges (spurs) crossing perpendicularly. Many streams emerge from a huge network of drainage channel in the hillslopes, which are mostly covered by forest and shrub, while terraces and flood plain are used for cultivation.

Geomorphological classification

The conceptual geomorphological classification for the Siwalik Hills in the Trijuga valley is shown in Fig. 2.9. It generally consists of four geomorphological zones: Hills, upper and lower terraces and floodplain. Based upon the aerial photo interpretation and field observations, the area can be divided into four geomorphological zones: Hill (>280m), Upper terrace (220-280m), Lower terrace (200-220m) and Flood plain

(<200m) (Fig.2.10). The classification is similar to the one done by Kimura (1997) as shown in Fig. 2.4.

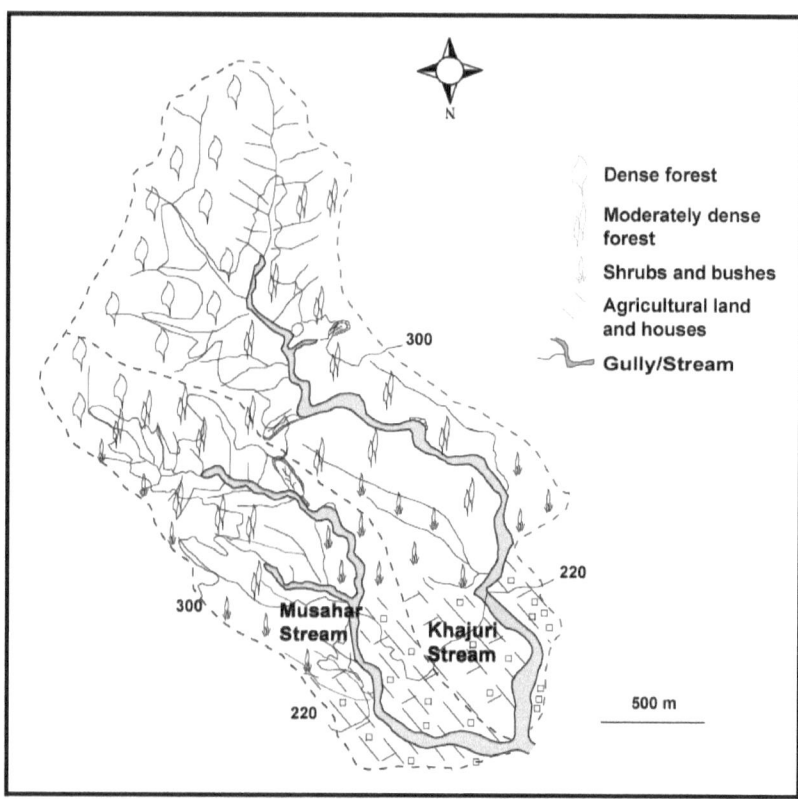

Figure 2.8 Topography, drainage and Land use in the Khajuri catchment.

The hill slopes are characterized by frequent instabilities despite thick vegetation cover. They exhibit low relief; however they are steep and dissected by many rills, gullies and landslides. The ridge and divides are generally sharp and narrow, which indicates the prevalence of rapid surface erosion. Most slopes are formed under the combined rather than the unique action of mass movement and running water.

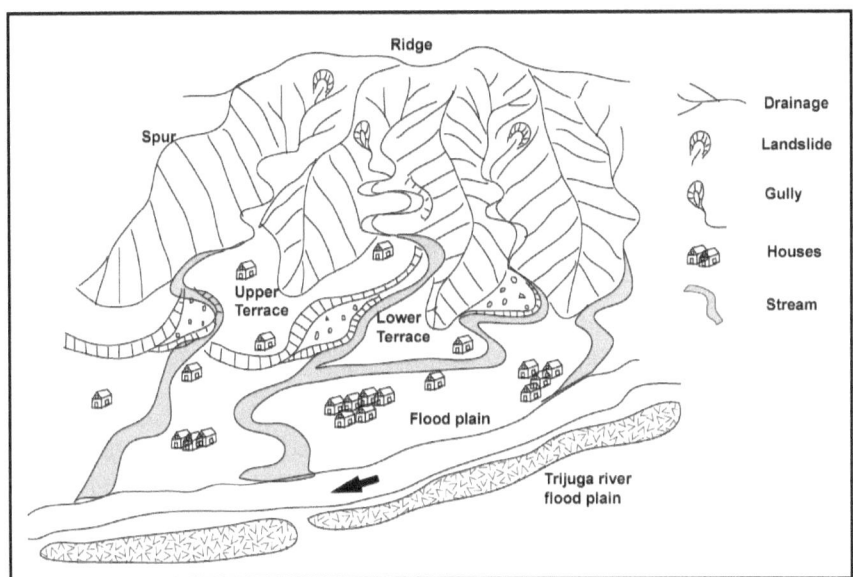

Figure 2.9 A sketch depicting the geomorphological units in the Siwalik Hills of Trijuga River valley.

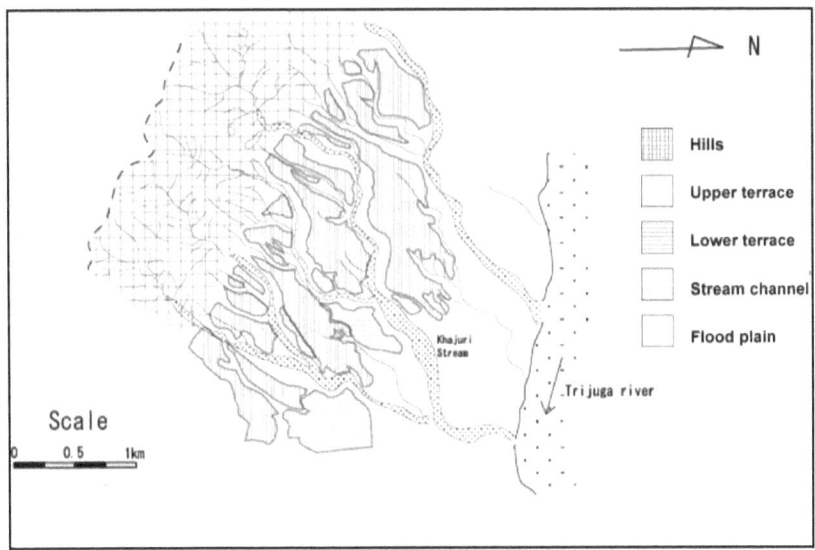

Figure 2.10 Geomorphological classification map of Khajuri watershed and surrounding area.

Drainage patterns

These two small streams are taken as the sites of various field measurements related to water and sediment. Khajuri is the main stream and Mushahar is its main tributary stream. These streams can be considered generally third order drainage channels.

The bed and bank materials control the erosive resistance, or erodibility of the channel boundaries. Hence, the channel formed on bed rocks (called confined channel) and on alluvium or sediment (called unconfined channel) exhibit different behaviour as regard to water and sediment mobilization. In this respect, it is important to note that the study streams are confined type mainly in the hillslope and terrace reaches. However, when they emerge from the terrace to the floodplain suddenly, they change from confined to unconfined channel. Hence, the geomorphological classification is the fundamental basis for the study of fluvial processes such as bank erosion and process of stream pattern change.

The form or morphology of the channel (including its size, cross-sectional shape, longitudinal profile and planform pattern) is the result of processes of sediment erosion, transport and deposition operating within the constraints imposed by the geology and terrain of the drainage basin (Thorne, 1997). Hence, the shape and structure of the drainage basin is important for the study of geomorphologic processes. Fig. 2.8 indicates that the two streams originate from the hills with a dense network of drainage channel. The shape of the drainage channel is somewhat parallel. Thorne (1997) mentions about drainage patterns and their geomorphic interpretation. A parallel pattern develops where there is a steep regional dip to the terrain that imposes a preferred direction of drainage. As the geological beds of the area generally dips towards north east (i.e. to the direction of main channel flow), the development of drainage channels seems to be strongly controlled by the geology of the area.

3
Research design, field measurement and techniques

Research design

As outlined in the objectives, the research is designed for the two spatial scales: one is related to the regional scale dealing with the issues such as changes in land cover and stream behaviour taking Trijuga valley as a case study area, and the other is related to intensive field measurement on a catchment scale, taking Khajuri stream as the study site. The research framework designed for the study is shown in Fig. 3.1.

The regional study can be considered as a background study for understanding key processes such as the historical trend in land cover changes and general behaviour of the streams. This understanding is worthwhile for gaining a more detailed and comprehensive understanding of these processes within a smaller spatial scale.

As defined in the research objectives, four types of hillslope erosion processes have been studied: surface, rill, gully and landslide erosion. The streambank erosion has also been studied in a greater detail to gain understanding of the in-channel erosion processes. Based on the field observation and measurement, the rate of sediment production from each erosion process (i.e. sources) is estimated. Estimation of sediment production from the entire catchment is then undertaken by summing up the sediment production volume from each individual source within the catchment. For this, results from plot-scale measurement have been extrapolated. The results should therefore be considered with caution when applying for any other geophysical environments.

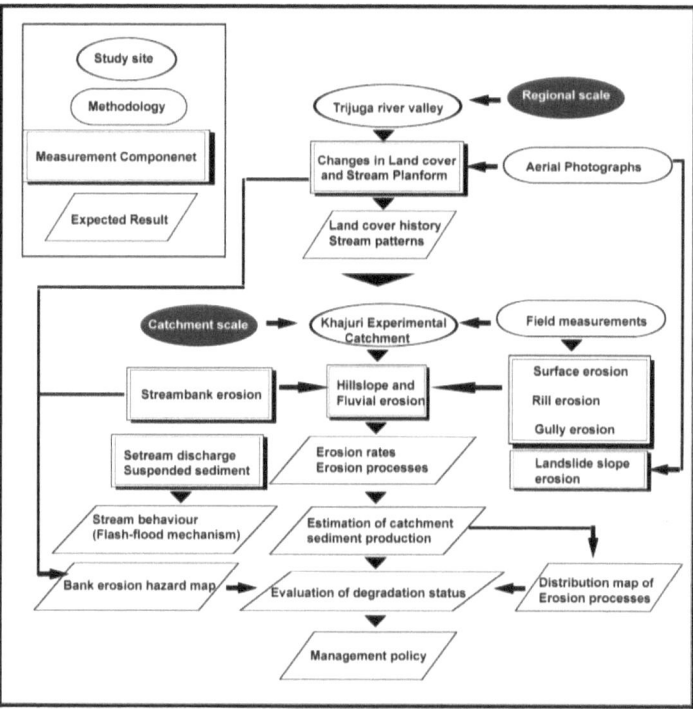

Figure 3.1 Research framework designed to identify and measure different geomorphological processes.

Two types of techniques have been developed and applied in order to assess the degradation status of the catchment: one is the *bank erosion hazard mapping* and the other is catchment-wide *erosion hazard mapping*. The bank erosion hazard map is prepared based on the field measurement of bank erosion and identification of stream bank characteristics. The hazard evaluation method is then compared with the characteristics of stream planform change. The erosion hazard map is produced based on the mapping of slope instabilities and field-based measurement of erosion rates. For this, a combination of methods has been adopted using both aerial photographs and field measurement. In particular, landslide distribution map has been developed based on the historical aerial photographs.

Field measurement

The locations of monitoring sites for different components of the study are shown in Fig. 3.2. The sites for the hillslope erosions are mostly chosen in the headwater reach. Most of the sites for bank erosion are located along the main stream of Khajuri. The sites were chosen bearing in mind that they are representatives of other areas within the catchment in terms of soil, land use, topography. The sites for the discharge and sediment measurement are located at the outlet of the catchment.

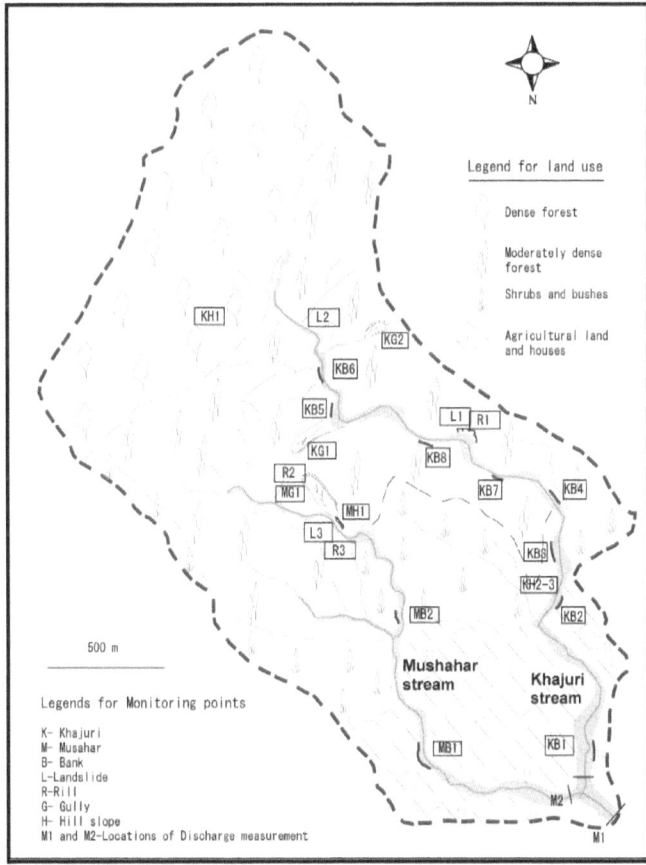

Figure 3.2 Monitoring network in the experimental catchment of Khajuri stream.

Techniques

In the regional scale study, the technique of aerial photography has been used for the assessment of land cover history and trend of stream pattern change. In the experimental catchment, field measurement is mainly includes the use of erosion pins. The two methods are briefly discussed here.

Study of aerial photographs

The study focuses on the use of computer-assisted interpretation of multi-temporal aerial photographs and satellite image. Black and white aerial photographs taken in 1964 (Scale 1:12,500) were obtained from Department of Forest, and that of 1978 and 1992 (Scale 1:50,000), were obtained from Department of Survey, Katmandu. Reference map was taken as the topographic map (1:25,000) compiled from the ground-verified 1992 aerial photograph, which was prepared by Land Resource Mapping Project (LRMP), Nepal. ASTER satellite image (resolution 15m, 2003 Nov.) was obtained from Earth Remote Sensing Data Analysis Center (ERSDAC), Japan.

First, the aerial photographs, satellite images and topographic maps were scanned with a resolution of 800 dots per inches and were saved in TIFF format. Then, using a computer software called ERDAS Imagine (version 8.5), the topographic map was georeferenced to the Universal Transverse Mercator (UTM) system by selecting four ground control points. Polynomial rectification of the 1992 aerial photo was performed using the same software by registering many ground control points (GCPs) on the topographic map as well as on the photo (Fig. 3.3). This process of rectification involves the stretching or compression of the image in as uniform a manner as possible in order to match the base map locations of ground control points. The 1992 rectified photo was subsequently used as reference image for the rectification of 2003 image and 1978 and 1964 aerial photographs in the same way. All rectified images were then imported to a software called- Adobe Illustrator version 10, where land use component and stream boundaries were traced and analysed by overlaying all rectified image layers with the topographic base map. Measurement on area of each land use component and stream cross-section was carried out in the same software. Interpretation and analysis of the aerial photographs as well as satellite image was done based on field observation and verification.

It is important to note that some errors were detected in the rectified images. Residual georeferencing error was estimated by fixing as many ground control points as possible leaving others "free" and then comparing actual and mapped location of the free points, the method followed by Micheli and Kirchner (2002). This procedure was repeated for various ground control points to generate spatially variable uncertainty estimates. Residual spatial error (difference between actual and mapped location of a feature) estimated for 2003 satellite image was ± 21 m and for 1992-photograph it was ± 7 m. This value is contrasted to errors of up to 47 m for satellite image and 35m for 1992 photograph estimated for non-rectified images of the same scale. The spatial errors for 1978 and 1964 images were ± 8 m and ± 5.5 m respectively.

In addition, there might be some errors in tracing the landuse boundaries, but working in a digital format allowed us to zoom in the features, which could limit the potential errors significantly. For all photographs, tracing was done by a single person who is well familiar with the actual field environment so as to keep the potential errors in tracing as minimum as possible.

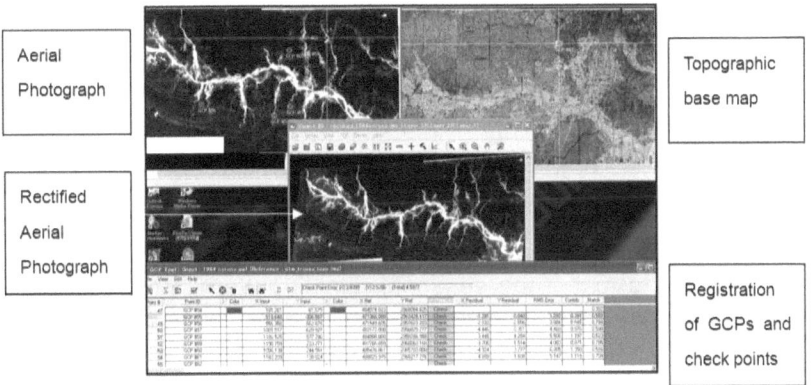

Figure 3.3 Geometrical rectification method using EARDAS Imagine.

Erosion pin method

This study in the experimental catchment is primarily based on the field monitoring works using erosion pins. Erosion pin method is simple, cheap and effective which can be applied for a wide range of field measurements of erosion. Erosion pin is made of iron nail with length ranging from 30 cm to 50 cm and diameter 8 mm to 12 mm.

For the measurement of surface erosion, the pins are inserted into the ground surface vertically with some 10cm exposed above it. Successive measurements in different time periods give the amount of erosion/deposition. For gully head erosion, a pair of pins is placed at each point of head-rim while for streambank erosion; the pins are placed perpendicular to the bank face. The method has been described in Fig. 3.4.

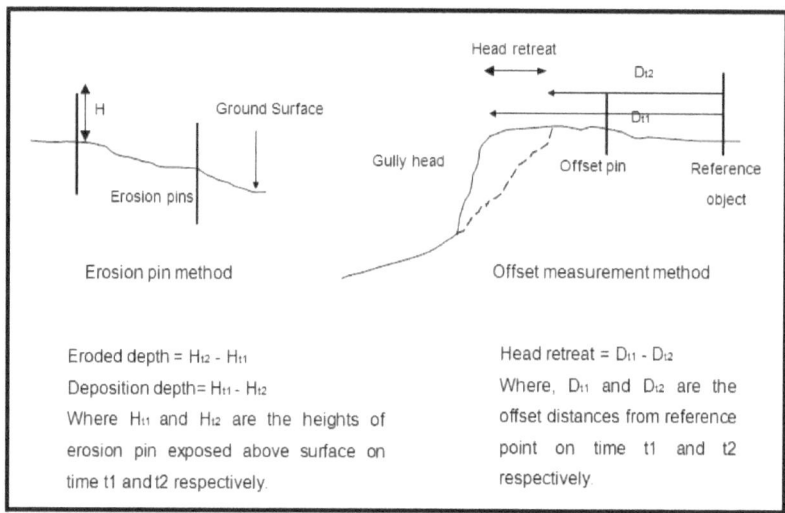

Figure 3.4 Erosion measurement using erosion pins.

Monitoring works were generally carried out three times a year in order to match the seasonal rainfall pattern. Monsoonal rainy season generally starts in June and ends in September. So, monitoring was done at least before and after the season, and also in other periods in case of extreme rainfall occurrence.

4
Changes in land cover and stream planform

Background

Changes in land cover and land use have important consequences for natural resources through their impacts on soil and water quality, biodiversity, and global climate systems (Houghton, 1994; Koning et al., 1999, Thomas, 2001). Land cover modification and conversion is driven by the interaction in space and time between bio-physical and human dimensions (Skole et al., 1994). As a human impact, many researchers have reported high rates of deforestation in developing countries and have debated their causes and consequences (eg. Tekle and Redlund, 2000, Pfaff, 1999, Tole 1998).

Use of forests and their products for different purposes has significantly changed forest cover during the last four decades in Nepal. Of the total 6.4 million ha of forests estimated in 1964, only 5.5 million ha of forest area was in natural stock. It has been estimated that a total of 0.1 million ha of forests in the Siwaliks and Terai were cleared under the government settlement programme from 1950s to 1985. An equal area was estimated to be lost during the same period due to illegal re-settlements. Overall, Nepal's forest area declined at an annual rate of 0.4 per cent during this period (HMG/ADB/FINNIDA, 1988).

Considering the forest estimates made during the LRMP (1978-1979), Master Plan (1985-86) and the 1994 inventory, the annual reduction rate of forest cover for the whole country is 1.7 per cent in between 1978/79 to 1994 and decrease in forest and shrub together is 0.5 per cent annually (MFSC, 1999).

Many studies have been conducted on landuse/land cover change indicating significant changes in forest cover as a result of human impacts. High rate of deforestation in the Middle Mountains is reported in many studies such as Zomer et al.(2001), Shrestha and Brown(1995), Stevens (1993) etc. while increase in forest cover has also been reported such as by Gautam et al.(2002) as a result of improved conservation strategy like community forestry. In the Siwalik Hills, Schweik et al. (1997) conducted a study on land-cover change in two small basins using aerial photographs and Geographic Information System (GIS). They found that deforestation was evident during the period between 1978 and 1992, indicating dense forest converted to maize-based agriculture in a basin while to degraded land in another.

In the present study, analysis on land cover change is undertaken first in the whole catchment of Trijuga river valley and then in the experimental catchment of Khajuri. Change pattern in the planform of the stream systems has been studied in the same order with the aim of understanding their morphological behaviour specifically width adjustment and shifting in course.

Land cover change in the Trijuga river valley

Four categories of land cover are distinguished: Forest, agricultural land, area of settlement and river/stream channels. The land cover categories were identified in the aerial photographs taken in 1964, 1978 and 1992, and in the satellite image taken in 2003. Results derived from the overlay of the land use categories are presented graphically in Fig. 4.1. Computation of temporal changes in land cover is shown in Fig. 4.2. The figure indicates that forest cover was decreased by about 17% (108.6 km^2) of total catchment area (639 km^2) during the period from 1964 through 2003. On the other hand, the area under cultivation increased by more than double. The main trade-off among the land cover types was the conversion of forest into cultivation land- a similar practice witnessed in other regions of Nepal and also in most developing countries in the world. The area under settlement increased significantly during the period 1978-1992. The area of stream channels was also increased by 23% from 1964 to 2003.

Temporal and spatial variations

The temporal and spatial change in land cover is found extremely variable. The deforestation, for example, was not uniform within the catchment (Fig. 4.3). Changes was maximum in the area- called Beltar to the north-east which is located adjacent to the confluence of Trijuga River and Koshi River, where rapid expansion of settlement and cultivation took place during the period between1964 and 1979. Many studies have attributed the cause of deforestation in the Terai region during this period to eradication of malaria. During that time many people migrated from the remote hill districts to settle in the densely forested lowland plains which led to massive deforestation (MOFSC, 1994).

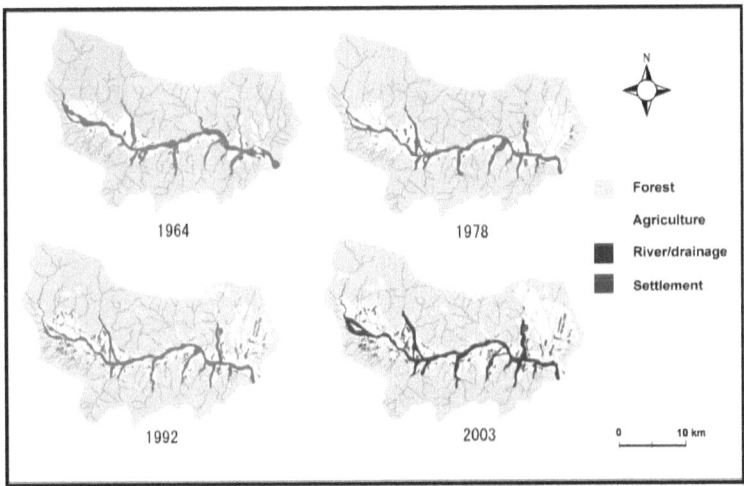

Figure 4.1. Changes in land cover from 1964 through 2003 in Trijuga river valley.

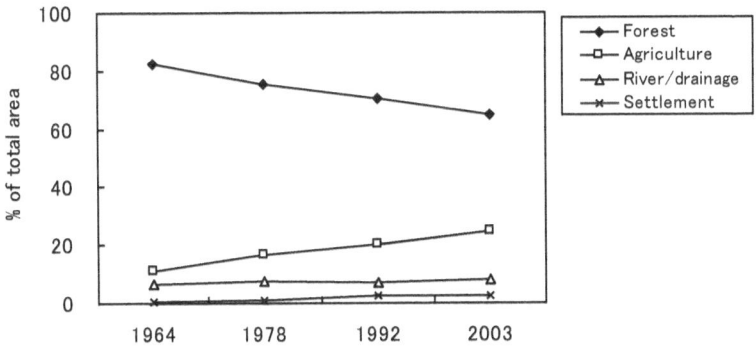

Figure 4.2 Computation of land cover change.

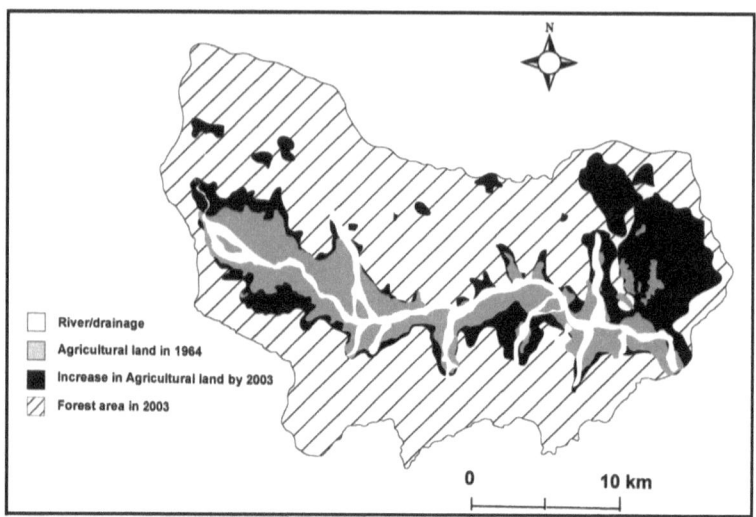

Figure 4.3 Deforestation occurred in the period from 1964 through 2003 in the Trijuga river valley.

During the period from 1979 through 1992, spatial pattern of land cover change differed markedly within the basin possibly due to the difference in geomorphology among other controlling factors. The outer Siwalik (Churiya) range (southern Hills from Trijuga River) consists of coarse sediments and boulder conglomerates whereas inner Churiya range in the north of the river is made up of less coarse sediments (Kimura, 1997). Consequently, the hillslopes in the south are not suitable for agriculture than in north. Hence, agricultural expansion and deforestation took place in a less pace in the southern hillslopes. From 1979 to 1992, there was significant increase in the area of settlement. There are main two reasons: first, a cement factory was constructed near the district headquarter-called "Gaighat" during the late 1980's which is the biggest one in Nepal. Second, the area was linked by a national highway which transformed the socio-economic condition of the whole basin.

During the period from 1992 through 2003, there was no considerable increase in the cultivation area especially in the southern region. This may be due to the fact that the floodplains and river terraces were already brought under cultivation by 1992. The expansion of cultivation into the hillslope is difficult as the soil is not generally favourable for agriculture. In addition, making terracing on the hillslopes is difficult as a result of steep topography. The rate of deforestation during this period is therefore found decreased overall. Another possible reason for this positive trend might be the practices of sustainable forest management such as community forest, the impact of which has been found positive in many other areas of Nepal (e.g. Gautam et al., 2002). In the present context community forestry is the most focused and prioritized program of the forestry sector in Nepal (DOF, 1995).

Land cover change in Khajuri catchment

In order to assess more comprehensively the pattern of land cover changes in the experimental catchment, similar technique of overlaying land cover categories was undertaken. The area is about 26.5 km^2 in area which consists of three small catchments including the experimental one.

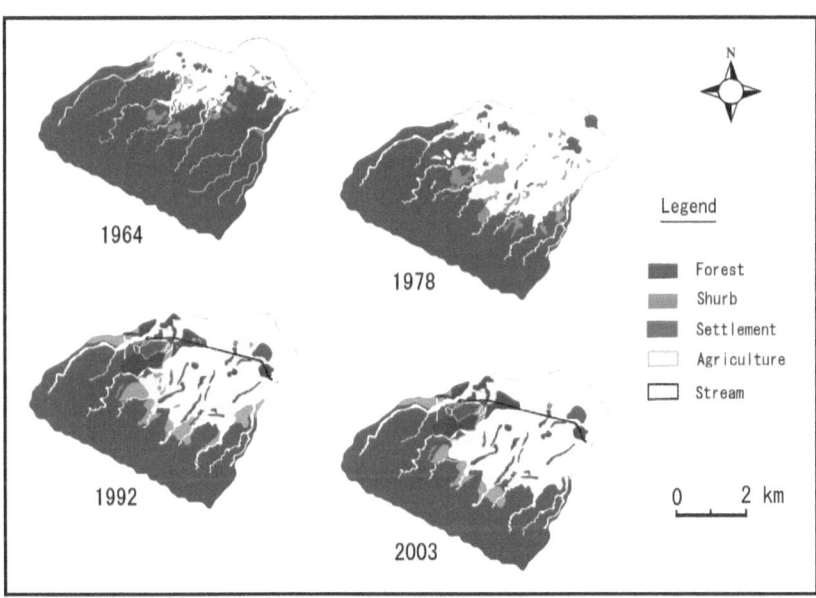

Figure 4.4 Land cover change from 1964 to 2003 in the vicinity of Khajuri watershed.

Five categories of land cover were distinguished: forested land, shrub land, cultivated land, area of settlement and stream channel. Overlay of the land use categories indicated that during the period between 1964 and 1978, forested land area decreased by 28% (21% of total study area) while agricultural land increased by 82% (15.5 % of total study area) (Fig. 4.4 and 4.5). More than 80% of forest loss was due to increased agriculture and remaining converted into other categories.

In the period between 1978 and 1992, both forest area and agricultural land decreased by 7.5 and 11 % respectively. Area of settlement increased remarkably by 300%. As stated earlier, the main reason for increase in settlement is the construction of the cement factory (Fig. 4.6). Like in the whole basin of Trijuga river, the forest area remained more or less unchanged during the period between 1992 and 2003 in the study catchment too.

In this area, forest management was transferred to the villagers under community forestry program that started in 1995 (DFO, 2003). A questionnaire survey conducted in the area in June 2003 indicated that majority of the people expressed satisfaction over the achievements from the community forests.

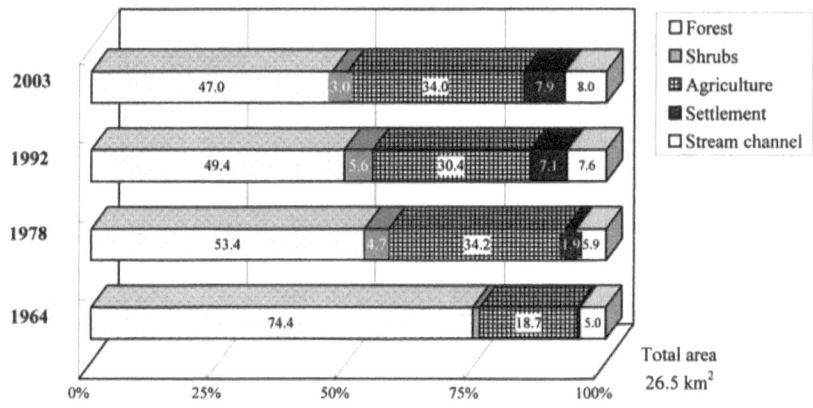

Figure 4.5. Computation of land cover change.

The most important finding relevant to this study is that the stream channels are widening progressively. Therefore, further study was carried out to examine the change pattern of stream channels and their potential consequences.

Analysis of stream planform change

It was noticed from the analysis of land cover change that stream area was constantly increasing over the recent decades, which can hint the on-going changes in morphological characteristics in the stream environment. The study of morphological changes can help to gain understanding of the stream behaviour in channel adjustment which in turn reflects the phenomenon of sediment mobilization by the processes such as bank erosion. So, an attempt has been made in this study to link the planform change behaviour to the process of bank erosion, especially in developing bank erosion hazard map which will be described later. The tributary streams of Trijuga river were analyzed for the changes in course. The location and characteristics of the study streams are presented in Chapter 2.

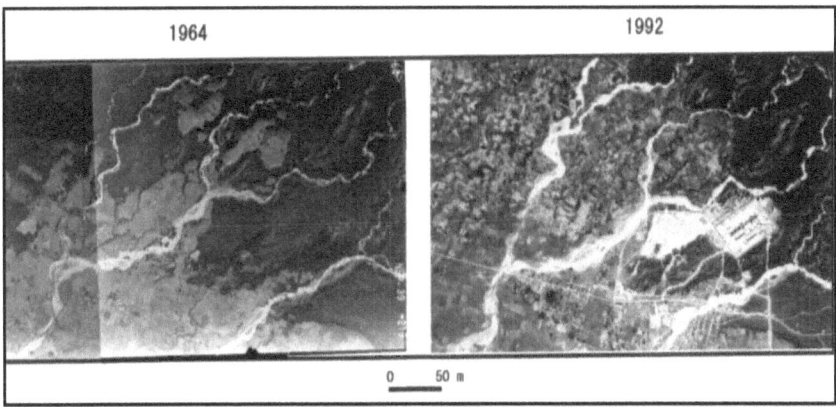

Figure 4.6 Comparison of aerial photographs taken in 1964 and 1992 in the Khajuri catchment. Significant increase in cultivation and decline in forest cover can be noticed. The expansion of settlement area was mainly due to the construction of a cement factory which is clearly visible on 1992 photograph. Also, there was a remarkable change in stream planform, mostly they widen around the floodplains.

In view of the difference in change pattern in different reaches, the streams were broadly divided into three reaches as mentioned in geomorphological classification: hill, terrace and flood plain. In the hills, stream channels are generally small and mostly confined by bedrocks. Hence we can not generally expect significant changes in the planform within the hill. In the terrace reach, the changes are relatively small which is difficult to be ascertained precisely from small-scale aerial photographs. On the other hand, changes in the most downstream flood plain, being larger in scale, can well be defined from the aerial photographs used in this study.

Overlay of stream boundaries traced from the rectified aerial photographs is shown in Fig. 4.7. It indicates that the streams are characterized by extremely movable boundaries with frequent widening, contraction and course shifting in the floodplain reach. Extent and pattern of stream width adjustment widely varies both in space and time.

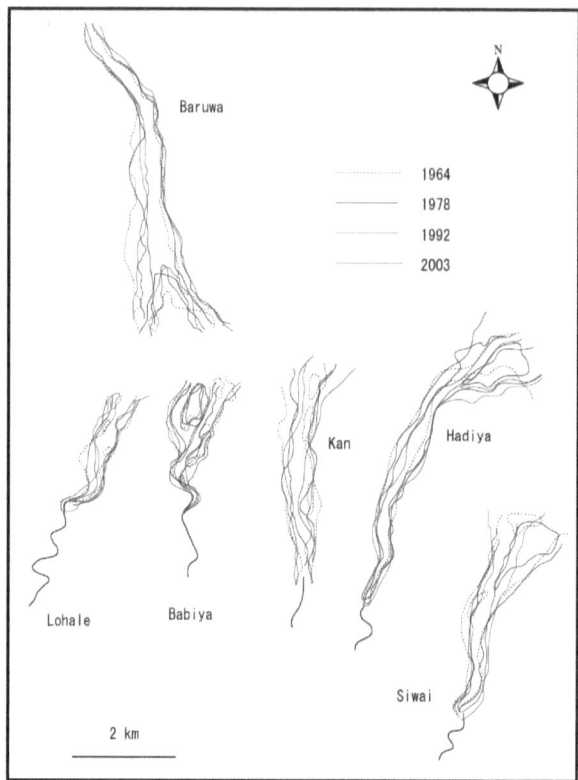

Figure 4.7 Delineation of planform change in the tributary streams of Trijuga River from 1964 to 2003.

The changes are more pronounced in the area of confluence i.e. area near the confluence of main river Trijuga. Course adjustment pattern of the Trijuga river may have the influence on the change pattern of the tributary streams near the confluence, however, this effect is not covered in the present study.

In general, the streams have undergone gradual widening in many locations mostly in the upper reaches of the floodplain. Course shifting is evident in most streams mainly in the middle and lower reaches of the floodplain. No segments of any stream are found perfectly stable. The amount of widening and shifting varies widely stream-to-stream and even within the stream. Maximum widening was found varying from a few meters to as high as 340 m.

Changes in planform in the experimental streams

As indicated by Fig. 4.6, the experimental streams Khajuri and Mushahar show clear changes in the planform during the study period. In most of the sections, they are characterized by increase in channel width. Fig. 4.8 depicts the overlay of the stream boundaries traced from rectified aerial and satellite images. Stream widths measured on the images for different time periods are shown in Fig. 4.9.

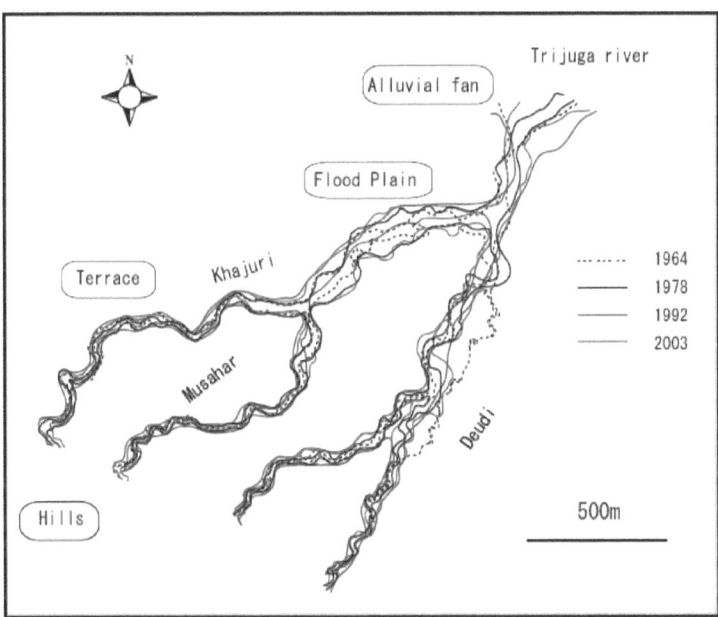

Figure 4.8 Delineation of stream planform changes in the experimental streams: Khajuri and Mushahar from 1964 to 2003 by aerial photo overlay.

In the downstream of the floodplain (confluence reach), not only stream widening is persistent but also there is channel course shifting phenomena. In the Khajuri-Deudi confluence, the stream seems to have shifted right side by about 200m. In the flood plain reach, the streams widened gradually. Some streams such as Deudi have straightened the meandering plan form. In the terrace reach also, the streams widened gradually.

It is important to note that in 2003, decrease in stream width is noticed predominantly in flood plain region. This is mainly due to the bank protection measures such as earthen embankments constructed by some agencies like the then Water Induced Disaster Prevention

Technical Centre (DPTC), a joint research initiative of Nepal Government and JICA, in the period between 1998 and 2000.

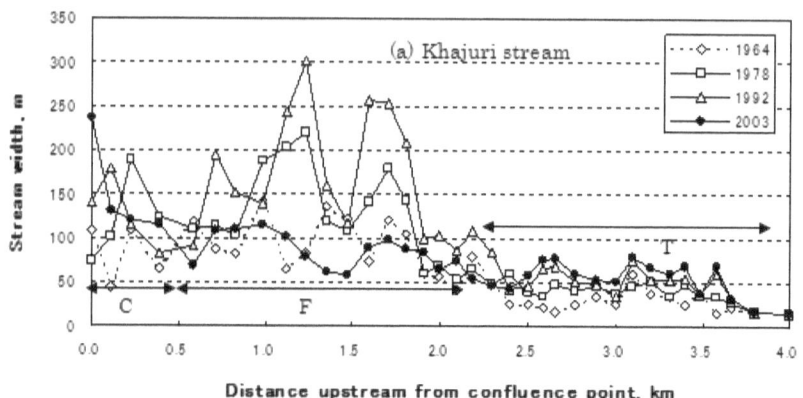

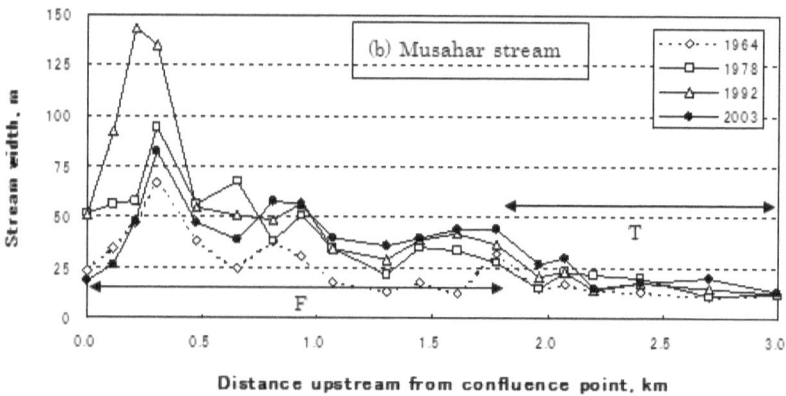

Figure 4.9 Stream width measured from aerial photographs (C- confluence, F- flood plain, T-Terrace)

Effect of land cover change on stream planform change

Effect of land use change is explained considering the foremost tradeoff between land use categories: conversion of forest into agricultural land. The hypothesis is that, when forest is converted into agriculture, runoff coefficient will be higher which could have triggered higher peak flows affecting the stream morphology. An attempt was made to test this hypothesis considering the area change in forest and agriculture and resulting change in stream planform.

However, there could not be found any conspicuous and meaningful trend that the changes in stream morphology are proportional to the changes in land use. This is clear from the fact that deforestation mostly occurred in the flood plains during 1964 to 1978, and in the terraces subsequently. By 2003, most of the area in the headwater had remained well covered by thick forest. However, morphological changes in the downstream are found going on in recent years even if rate of deforestation (i.e. conversion of forest into agriculture) is diminished. Although further investigation is required to test the hypothesis in the whole area, no clear effect of land cover change has been noticed from this study.

Instead, one possible effect of bank vegetation can be of interest because some changes were detected over the period of four decades. So, an attempt was made to examine the change pattern in relation to the change in bank vegetation in some streams. The hypothesis is that near-bank vegetation generally strengthens the stream banks resulting in the stable channel.

Three categories of vegetation were defined: forest, shrub and bare. Fig. 4.10 depicts the changes in bank vegetation detected in the experimental stream. In the period between 1964 and 1978, bank vegetation was found destroyed in some locations of the floodplain. Conversion of forest into shrubs was also noticed in some banks. From 1978 to 1992, the conversion continued upstream. In the period of 1992 to 2003, clearance of shrubs was noticed in some locations. Engineering protection works such as embankment were carried out along the banks in the floodplain during this period.

When comparing the vegetation change map (Fig. 4.10) with the course change pattern (Fig. 4.8), again no clear effect of changes in vegetation was noticed on the change pattern of the stream. This is clear from the fact that remarkable changes occurred also in the locations where bank vegetation was not changed. In the floodplain, where no bank vegetation was detected in 1964, remarkable widening and course shifting was still noticed following that period as mentioned already. Similar qualitative comparison was carried out in the other streams too, but clear effect of bank vegetation could not be noticed. More detail analysis on the effect of bank vegetation is given under streambank erosion later.

Stream planform change process

Field observation and measurements indicated that a complex interaction takes place between bank surface and stream runoff flow. Expansion of stream section through the process of bank erosion depends on many factors such as slope, bank material, bank geometry and near-bank vegetation etc. Banks are composed of alternate layers of fine and coarse-grained soil material with occasional layers of boulders. Process of selective scouring takes place, which is an intermittent process and very intricate in nature. In many instances, a recovery of stream section is also noticed. It is mainly due to the protection by fallen debris from high banks. Gradual development of vegetation on such debris slopes protects further bank scouring. Comprehensive analysis on the process of bank erosion has been done under section of streambank erosion.

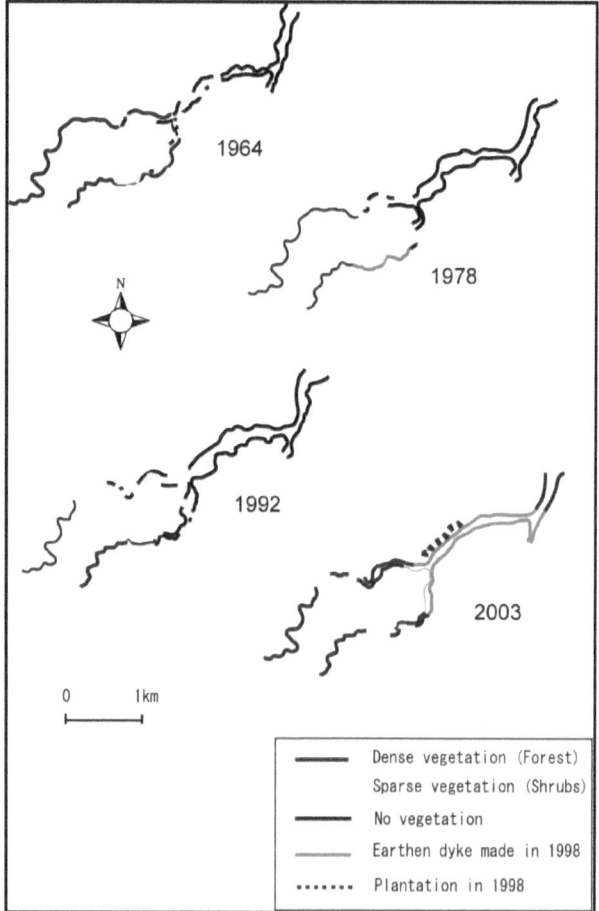

Figure 4.10 Changes in bank vegetation in Khajuri stream.

Another process is the cutoff formation in the meandering points, however thresholds to initiate this process are less understood. In the flood plains, lateral inundation is the main process, which occurs mainly during torrential downpours. In such condition, quick and transient peak flow discharge is generated which is often higher than bank full discharge.

An interesting mode of planform change has been noticed in the floodplain, which is under extensive cultivation. Riparian farmers keep on their efforts to protect their land from the frequently shifting stream channels using various techniques. As a result, the channel course may remain stable unless a big rainfall event occurs. In many streams, contraction of channel widths is also observed because of such protection efforts (Fig. 4.11). Since the protection techniques are of temporary types (mostly earthen), large floods can easily destroy them, which again may result in stream widening or course shifting.

Figure 4.11 Protection of agricultural land by earthen bunds.

Driving factors of stream pattern change

Channel pattern represents a mode of channel form adjustment in the horizontal plane. It is an additional mode of channel form but linked closely with transverse (cross-sectional shape and size) and lengthwise (channel bedslope) modes. It influences the resistance to flow and can be regarded as an alternative to slope adjustment when valley slope is treated as constant at the short and medium timescales.

Channel form adjustment processes largely depend on the time scales. Four types of time period can be defined (Knighton, 1998): instantaneous time (< 10 years), short (10-100 years), medium (100-1000years) and long (>10000 years). Channel forms are not simply the product of the instantaneous conditions but of the processes that operate for a long time. The short and medium time scales are the most relevant as regards channel form adjustment since mean water and sediment discharge are independent variables to which an average channel geometry is related (Knighton, 1998).

The results indicated that during the short time scale of four decades the study streams exhibit remarkable changes on their pattern. It is important to note that the case of stream width reduction by natural process is evident in very few locations compared to the expansion or course shifting. It may reveal the fact that the streams are still in the phase of development and are yet to show the characteristics of stable geometry.

The ability of the channel to shift laterally depends on the resistivity of the banks (Hickin and Nanson, 1984). Channels remain straight if little or no erosion occurs, whereas meandering requires localized bank erosion. Hence, because of relatively steep local relief and unconsolidated bank materials, localized bank erosion widely persists especially in the upstream reaches of the Siwalik Hills.

The river channels are one of the most sensitive components of the physical landscape, with an ability to respond rapidly to disturbance. Disturbances to external factors such as climate and vegetation are a major cause of instability in the fluvial system. However, when analyzing the rainfall data in the study area, no clear evidence of climate change can be noticed at least over the periods of last four decades (See Chapter 5). Vegetation condition also remains more or less unchanged in the headwater reach although deforestation was evident mostly in the lowland floodplains. Therefore, it is likely that the changes may arise due to the in-stream processes rather than any external disturbances. Schumm (1973) asserted that change can also be initiated internally during the continued operation of prevailing input levels in the fluvial system. Patton and Schumm (1981) suggested that, in regions with high sediment yields and high ratios of sediment discharge to water discharge, episodes of channel cutting and filling can occur under relatively stable climatic condition because of the control exerted by geomorphic parameters that are intrinsic to the system, notably those that are slope-related.

The dominant geomorphic parameter, however, is difficult to define specifically in the case of the study streams. There are many controlling factors affecting discharge and sediment transport which influence channel form at a variety of scales. Adjustment to the internal geometry of the fluvial system involve a large number of variables whose interdependence is always not clear because the role of a single variable can not easily be isolated (Knighton, 1998). Hence, the study of long-term changes in hydrologic and sediment transport characteristics can help answer partly the mystery of channel change within the Siwalik Hills.

Implications of land cover and stream morphological changes

Vegetation cover is one of the primary controls on catchment hydrology and sediment supply, and is the control most susceptible to human disturbance. A major consequence of forest clearance is accelerated soil erosion on hillslopes, associated with which are network extension through gully development and an increase in the amount of sediment supplied to the streams (Knighton, 1998). The conversion of forest into agricultural lands also affects the runoff characteristics. As well as affecting erosion rates on hillslopes, the increased runoff can accelerate bank erosion along tributaries and floodplain sedimentation throughout the drainage system. However, in the case of the study catchment, other controlling factors could be more important than the land cover change in producing generally widening behaviour of the streams.

Clear difference in the erosion rates from the forested and bare land reveals the effect of land cover on erosion. This and other many studies have verified the role of vegetation cover in lessening accelerated erosion. Hence, encouraging vegetation cover in the degraded slopes of the Siwalik Hills should certainly be a management target.

There could be two principle implications of geomorphic changes of stream systems. First, agricultural land in the terraces and flood plain area are damaged by bank erosion and inundation. Second, the streams sometimes destroy areas of settlement, mainly in flood plain inflicting loss of lives and property, which will be discussed later.

Flood control countermeasures such as spurs and embankments and bank protection works such as Gabion walls and riparian vegetation etc. could be appropriate in controlling the mobile boundaries of Siwalik streams. This is also evident by the results of this study as the treated streams have exhibited stable and defined boundaries as a result of applied countermeasures. From field observation, it was noticed that in such streams like Khajuri and Musahar, significant area of flood plain has been reclaimed, thanks to the effectiveness of earthen embankments. The embankments which were constructed by DPTC/JICA in 1998 (see Chapter 11) are found effective in protecting the adjacent land from flooding. However, in many other similar streams, the problems are still severe.

The adverse implications of the geomorphic changes of Siwalik streams have pointed out the urgent need to control them. Many flood plain protection works are being carried out in an indiscriminate manner mostly as emergency relief but coordinated planning for hazard mitigation in regional and local scales is lacking. Conservation measures must be effective in terms of function as well as cost. The measures therefore can not be applied on the whole reach of the streams or all streams at a time. Hence it is necessary to identify and prioritise the most vulnerable stream locations for planning of protection structures. Bank erosion hazard mapping is a useful tool in this context which has been discussed in chapter 9.

Summary

Following points can be summarised:

- Land cover change analysis shows that main pattern of land use change in the Siwalik headwaters was encroachment of forest for expanding agriculture, a similar trend on other parts of Nepal. However, in the recent past the rate of deforestation has been diminished possibly due to unsuitability of hillslopes for agriculture as well as positive outcomes of community forestry program.
- The streams have undergone significant changes in planform pattern. The extent of change is more pronounced in the flood plain and confluence areas. Most of the streams or most of the sections in a stream indicate the increase in width over the period of four decades. The increase in width varies from few meters to up to 340m.
- No clear evidence was noticed on the direct effect of land cover change on the pattern of stream planform change.
- Some minor changes were detected on the streambank vegetation, forest being changed into shrubland in some locations. However, that could not be related to the stability of the stream channels. It is likely that some other control factors such as bank

material may have an additional effect on planform change process than the vegetation. It has been discussed in chapter 9.

- The cause of channel changes seems to be related to in-stream geomorphic controls rather than to external ones, however dominant control factors are difficult to identify within the scope of this study.

- Planning of stream protection works should aim to identify the changing pattern of the streams. Overlay of the stream planform using historical aerial photographs like done in this study can be a useful tool to identify and priority the control works at the most vulnerable locations. Likewise, bank erosion hazard mapping is another tool that can be used to protect the streambanks in priority basis depending upon the level of hazard.

5
Rainfall characteristics

Rainfall trend in Nepal

Nepal is primarily under the influence of the southwest monsoon, which is the rain bearing winds flowing from the southeast Bay of Bengal. Rainfall distribution varies greatly due to topographical diversities, but most of the areas show a distinct summer peak generally from June through September, and a prolonged dry season from October to May. The long-term mean precipitation in Nepal shows a decreasing trend from east to west (Chalise et al., 1996). Highest annual precipitation of up to 5000 mm occurs in the region of Pokhara in Kaski district (Middle Hills) and driest part of the country is the rainshadow area of Mustang on the Tibetan plateau with below 200 mm annual rainfall. Intense rainfall occurs generally on the foothills of the Siwalik Hills.

Rainfall data from period 1971 through 2003 were analyzed to gain understanding of any trend and pattern in the rainfall. The rainfall data were obtained from Department of Hydrology and Meteorology (DHM), Kathmandu. An observation of the data indicated that the data were missing for the period 1998-2000. The rainfall data of 2004 (only for rainy season) were obtained from the rainguage installed in the study area. Since the rainfall data of 2004 were not complete, they are not considered to estimate annual average rainfall for the study area.

Annual and seasonal rainfall pattern

The study area of Trijuga valley receives about 1880 mm of rain each year in average (Fig. 5.1). Table 5.1 shows the yearly rainfall totals in comparison to the long-term average rainfall for the whole period and the rainfalls during the rainy season of June to September. The table indicates that the area received highest rainfall of 3234 mm in the year 1974 and lowest of 931 mm in 1994. It also shows that when compared to the long-term average rainfall, the study years of 2002 and 2003 comprise 92 and 85% respectively. If looked into the rainfall pattern in rainy season, 70 to 90 % of annual rainfall generally occurs in this season. Looking at these figures and those of some extreme events, the rainfalls during the study years can be considered normal events which can be considered representative of the long-term average trend of the rainfalls even though inter-annual variation was considerable with coefficient of variation (CV) of 0.24 (Table 5.2) .

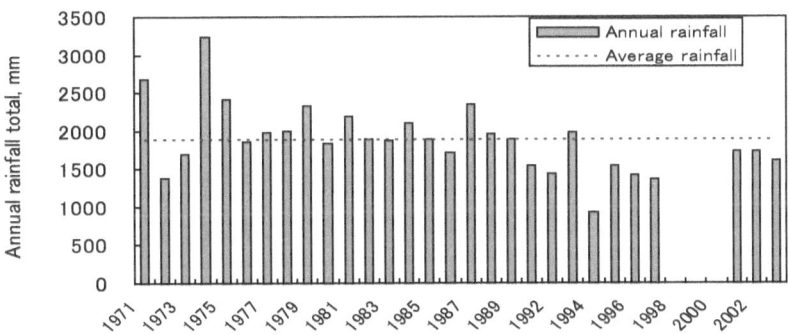

Figure 5.1. Annual and average rainfalls (DHM, 2003)

In view of the variation in the seasonal and monthly pattern of rainfall, and potential variation in the runoff and sediment production, monthly total rainfalls were analyzed. Generally rainfall starts in early period of June and ends at the late September. Considering this rainfall pattern, variation in the monthly rainfalls was examined for the whole period which is shown in Fig. 5.2. The figure indicates that rainfalls in the months of May and September were usually lower in comparison to other months for most of the years. Generally rainfall peaks are observed in the months of July and August. In the study years 2002 and 2003 also, rainfalls were maximum in the month of July.

Monthly rainfalls were averaged out for the years 1971-2001 and then compared with the monthly rainfalls for 2002-2004 (Fig. 5.3).

Rainfall in 2003 had almost average rainfall in the months June and July whereas lower in August. However, rainfall in 2002 had more rainfalls in July than in average, with less in the rest months. Thus, compared to 2003, rainfalls in 2002 had more uneven distribution in the rainy season. Rainfall in 2004 gained much higher peak in the month of July, however, data were incomplete for August.

Table 5.1. Comparison of rainfalls of study years (2002 and 2003) with the historical rainfalls. Station- Udayapur Gadhi (DHM, 2003). Here, rainy season implies the period from June to September.

Year	Rainfall mm	% of average rainfall	Rainfall in rainy season	% of annual rainfall	Year	Rainfall mm	% of average rainfall	Rainfall in rainy season	% of annual rainfall
1971	2671	142	1734	65	1986	1707	91	1307	77
1972	1379	73	1209	88	1987	2352	125	1888	80
1973	1696	90	1329	78	1988	1953	104	1400	72
1974	3234	172	2557	79	1989	1892	101	1366	72
1975	2413	128	1993	83	1991	1547	82	1385	90
1976	1858	99	1519	82	1992	1435	76	1098	77
1977	1977	105	1465	74	1993	1983	106	1588	80
1978	1993	106	1533	77	1994	931	50	746	80
1979	2333	124	1910	82	1995	1533	82	1316	86
1980	1831	97	1337	73	1996	1410	75	1267	90
1981	2183	116	1589	73	1997	1370	73	1176	86
1982	1897	101	1455	77	2001	1732	92	1309	76
1983	1877	100	1361	73	**2002**	**1727**	**92**	**1305**	**76**
1984	2092	111	1894	91	**2003**	**1604**	**85**	**1195**	**75**
1985	1897	101	1580	83					

Table 5. 2 Statistics of the annual rainfalls.

Parameter	Value
Mean	1880
SD	447
Min	931
Max	3234
Range	2303
CV	0.24

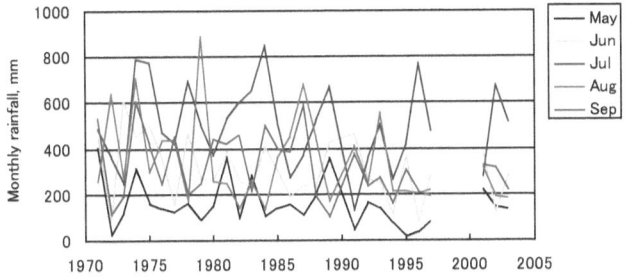

Figure 5.2. Comparison of monthly rainfalls during the rainy season from May through September.

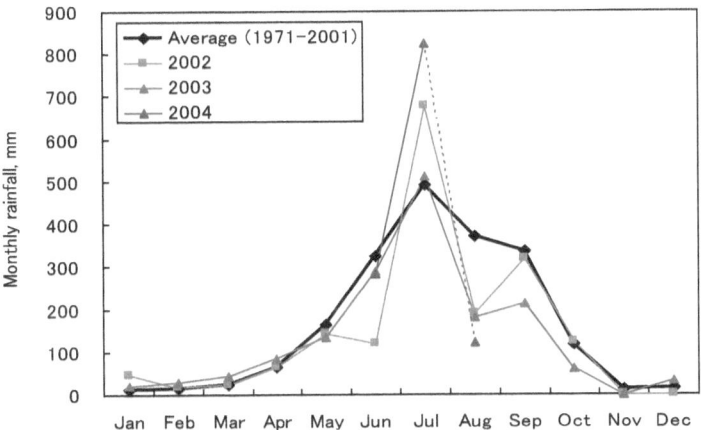

Figure 5.3 Comparison of monthly rainfalls during the study period with historical data.

Daily rainfall pattern

Daily rainfalls were analyzed for their size and distribution patterns. Fig. 5.4 shows the daily rainfalls in the study period 2002-2004. As mentioned earlier, daily data were available only in the rainy season of 2004. The figure shows that there was not much variation in the rainfall patterns of years 2002 and 2003 compared to of 2004. Fig. 5.5 indicates the frequency distribution of the daily rainfalls in 2002 and 2003, which implies very similar pattern. No rainfall more than 50 mm a day occurred during the years. In contrast, daily events in 2004 rainy season are much intense and larger. Of them, two events reached more than 160mm while four other events were more than 50 mm.

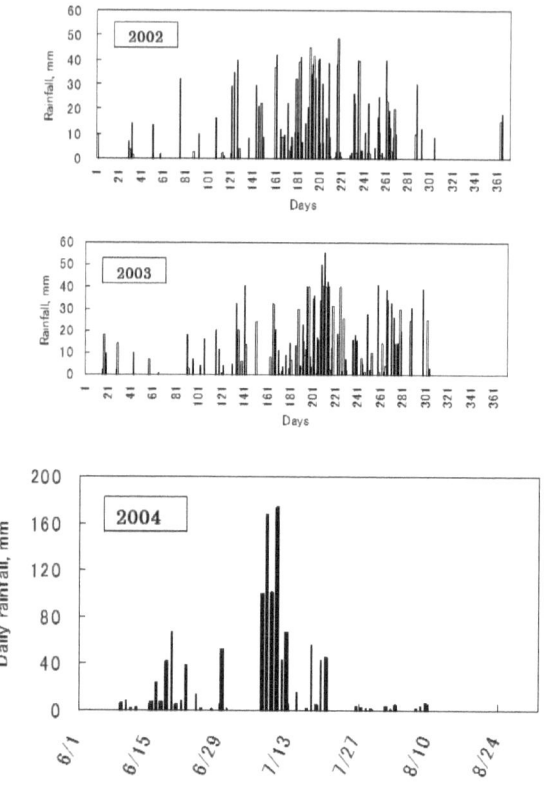

Figure 5.4 Daily rainfalls in the study years 2002, 2003 and 2004

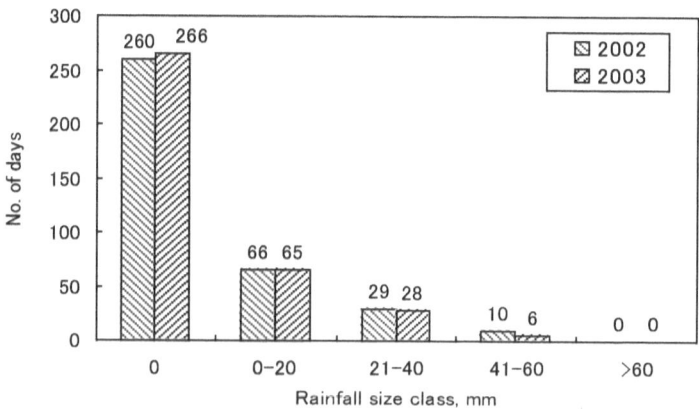

Figure 5.5. Frequency distribution of daily rainfalls in 2002 and 2003.

Major rainfall events

Since rainfall data on daily basis were only available since 1992, major rainfalls have been identified for the period since then. Some of the major daily rainfall events are shown in Fig. 5.6. The figure shows that at least one such rainfall occurs in most of the years. The ever-biggest event in the study area was on 13th August, 1995 which recorded 222 mm of rainfall. Here it is worthy to mention that this was the rainfall event, worst during the recent past of 50 years or so in terms of effect to the people and damages to the local infrastructures. It was confirmed in the field by interviewing with the local residents. Compared to the past events, occurrence of big rainfalls in 2004 seems much higher.

Frequency analysis and return period

The objective of frequency analysis of hydrologic data is to relate the magnitude of extreme events to their frequency of occurrence through the use of probability distributions. In this concern, return periods of annual rainfall totals were computed using three different frequency distributions namely Gumbel, Pearson Type-III and Normal distribution.

Frequency analysis has been done by using the method developed by Chow (1951). General equation for hydrologic frequency is given by:

$$X_T = \mu \pm \sigma k \quad \quad \text{Equation (5.1)}$$

Where, X_T is magnitude of an extreme event; μ is average of sample; σ is standard deviation and k is frequency factor. Frequency factor k is calculated by using standard equations of each distribution. Detailed equations for calculating the frequency factor (k) are given in Chow et al. (1988). Frequency factors and rainfall totals for different return periods are shown in Table 5.3, while comparison of the three type of distribution is shown in Fig. 5.7.

Fig. 5.7 reveals that the three types of distribution varied widely especially for longer return period. Estimation by Gumbel distribution seems higher, which is generally suitable for extreme type of events. However, the Pearson Type-III distribution seems to be in average of the two and therefore it is recommended to use in the present study.

The most important objective of the rainfall analysis is to evaluate the rainfall patterns during the study period comparing with historical data. As mentioned earlier, on the basis of annual totals, the observation years 2002 and 2003 can fairly be taken as normal and representative of the long-term rainfall patterns in the study area. However, rainfall patterns in 2004 (during the rainy season) seem quite higher than average, as some exceptionally high rainstorms were recorded in the month of July.

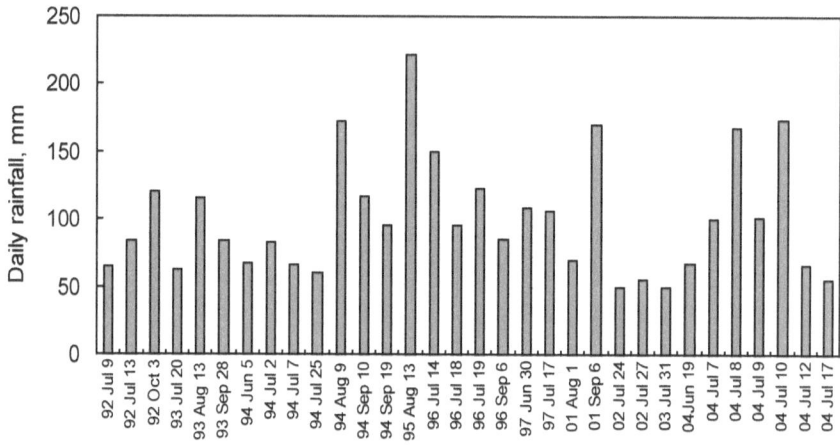

Figure 5.6 Major rainfall events (greater than 50 mm).

Table 5.3 Frequency factor and return period of annual rainfalls.

Return period, yrs	Gumbel distribution.		Pearson Type-III		Normal distribution.	
	k	Annual rainfall mm	k	Annual rainfall mm	k	Annual rainfall mm
5	0.867	2267	0.746	2213	0.842	2256
10	1.543	2569	1.341	2479	1.282	2453
25	2.390	2948	2.064	2802	1.751	2662
50	3.030	3234	2.583	3034	2.054	2798
100	3.660	3516	3.085	3259	2.326	2919

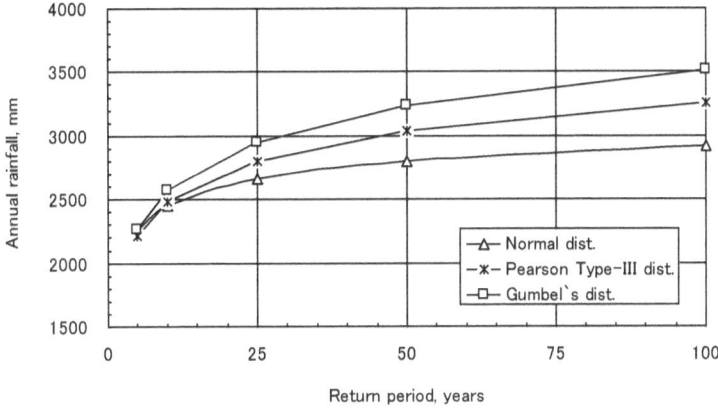

Figure 5.7 Return period of annual rainfalls.

From the analysis of frequency distribution, it is obvious that return periods of the annual totals of 2002 and 2003 are below 5 years. Hence, from this viewpoint also, they can be considered quite frequent and fairly representative. Caution must be taken however, when analyzing the data of 2004.

Summary

Following points can be summarised:
- The study area receives about 1880 mm of rain each year in average.
- About 70 to 90% of the total annual rainfall occurs during the rainy season from June through September.
- Generally rainfall peaks are observed in the months of July and August. In the study years 2002 and 2003 also, rainfalls were maximum in the month of July.
- The return periods of the annual totals of 2002 and 2003 are below 5 years.
- On the basis of annual totals, the observation years 2002 and 2003 can fairly be taken as normal and representative of the long-term rainfall patterns in the study area. However, rainfall in 2004 seems more intense and bigger.

6
Hydrology, flood and sediment transport

Background

Flood is the widespread hazard problem in Nepal. From the valley bottom in the Mountain Hills to the flood plain of Terai, flood is the regular threat to many lives and properties every year in the rainy season. Severity of these floods in terms of damages is widely documented; however, flood-triggering mechanisms are still the subject of discussion.

The floods in the streams of Siwalik Hills are typical and entirely different from the general floods in the Mountain Rivers. The most important factors in this context seem to be related to scale and geomorphological environment. Obviously the streams are smaller in scale. They cross the high relief hillslopes travelling a relatively short distance which result in the high velocity of stream flood and erosive power. The people know the streams for the high erosivity and uncertainty, the potential effects are generally out of their imaginations.

Measurement of flood

Rainfall and runoff measurements were carried out in the two study streams- Khajuri and Mushahar in 2002 and 2003. The key characteristics of the streams are presented in Table 6.1. Gauging stations were set up to measure the flow discharge at relatively stable sections of the streams (Fig. 3.2, Chapter 3). The flow velocity was measured by using floating method, and discharge was computed by establishing a stage-discharge relationship for the stream.

Flash floods have a high level of event independence since they are generally associated with isolated thunderstorms over small basins (Knighton et al., 2001). Hence, an event-based approach provides the most appropriate means for analyzing the hydrology of Siwalik streams. Fig.6.1 shows the hydrographs of some selected flood events measured in 2002.

Table 6.1 Characteristics of the study streams.

Stream name	Catchment area (km^2)	Stream length (km)	Av. Slope %	Drainage density (km/km^2)
Khajuri	2.84	3.68	2.0	3.52
Musahar	1.78	2.63	2.5	5.49

There were 7 rainfall events, which produced significant runoff in both streams in the period between the mid-June and mid-July, 2002, i.e. the peak of rainfall season. All the events except event 5 were single types, producing one well-defined peak discharge to which there is a progressive rise from zero and after which there is a progressive return to zero. Characteristics of the rainfall events are given in Table 4.5.

There are mainly three characteristics represented by the flood hydrographs to be discussed in this context - event magnitude, event duration and time to peak. Event magnitude is represented by two variables- total volume (V_t) and peak discharge (Q_p). The highest total volumes of flow measured were 58 x10^3 and 31 x 10^3 m^3 in Khajuri and Musahar streams respectively. Both were produced by the rainfall of 4th July (event 5).

Another variable- peak flow is often taken as an important variable in the flash flood as it determines the threshold height of maximum floods in reference to the topographical setting of the surrounding area. The highest peak flows measured were 9.5 and 7.2 m^3/s in the Khajuri and Musahar streams respectively as a result of event 3. Time to peak flow has typical characteristics. It occurs within a very short time period: within 30 to 40 minutes from the beginning of flood.

Duration of the flood obviously depends upon the duration of rainfall, and watershed characteristics. Flood hydrographs indicate that the duration often varies from 1 to 5 hours. For single events, most of the flows almost vanish within 2 hours.

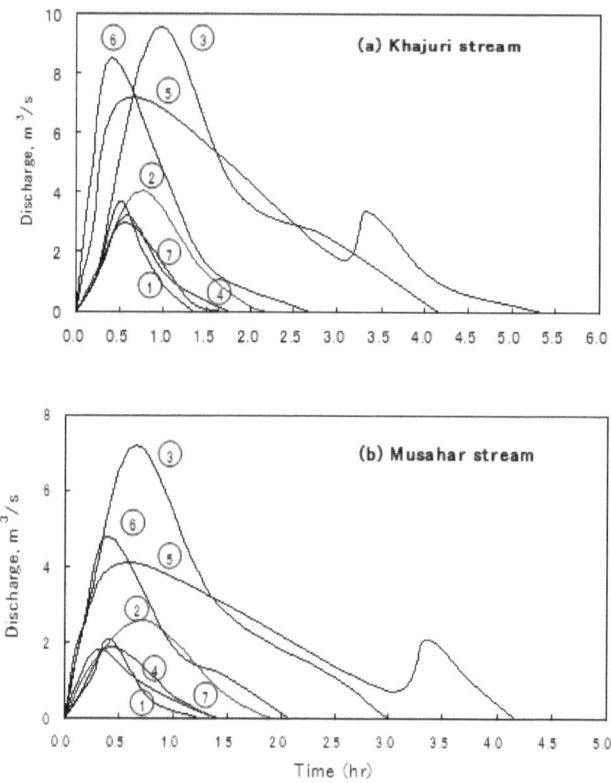

Figure 6.1 Flood hydrographs for some selected rainfall events.

Table 6.2 Characteristics of the rainfall events (corresponding to the hydrographs given in Fig. 6.1).

Event no.	1	2	3	4	5	6	7
Date	6, June	11, June	30, June	1, July	4, July	6, July	9, July
Rainfall, mm	13	17	45	19	31	37	17
Imax, mm/hr	19	30	40	31	29	34	41

Rainfall-runoff relationship

Rainfall-runoff relationship indicates an increasing trend of runoff with increase in rainfall. Accordingly, runoff coefficient (C) varies from almost 0 to 0.72 depending on the rainfall characteristics. Higher the rainfall, higher will be the runoff coefficient in general. The peak flow tends to vary linearly to the size of rainfall. It is partly because of the typical characteristics of the rainfall events that higher rainfalls were characterized by the higher maximum intensities. Rainfall-runoff relationships are shown in Figures 6.2 and 6.3 and rainfall-peak runoff correlation in Fig. 6.4.

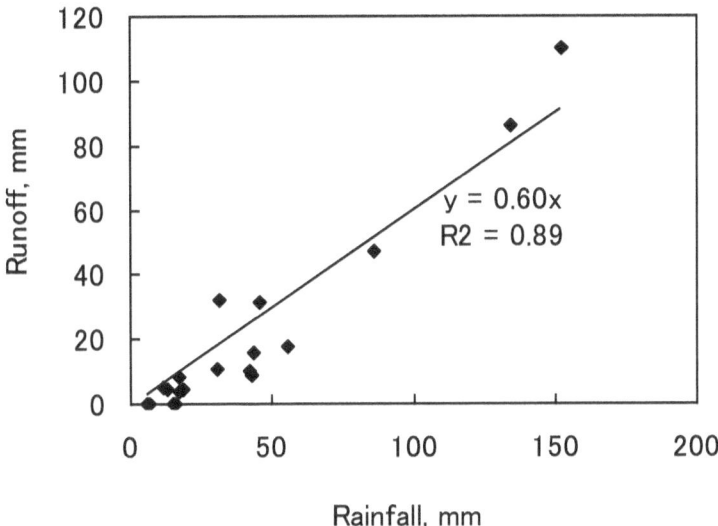

Figure 6.2 Relationship between rainfall and runoff in Khajuri main

The Siwalik streams are characterized by a very quick generation of peak flood as in case of the study watersheds whereby the time to peak is in the range of 40 minutes from the start of the runoff. This is due to the fact that it is governed mainly by the travel time of the flow from the farthest point in a catchment, hence shape and drainage density of the watershed. As there is a vast network of steep rills and gullies in the headwater zone, they serve for a quick disposal of channel flow. Since they begin just a few meters down from the ridgeline, the conversion of overland flow into channel flow takes place very swiftly once the generation of overland flow onsets (Ghimire et al., 2003).

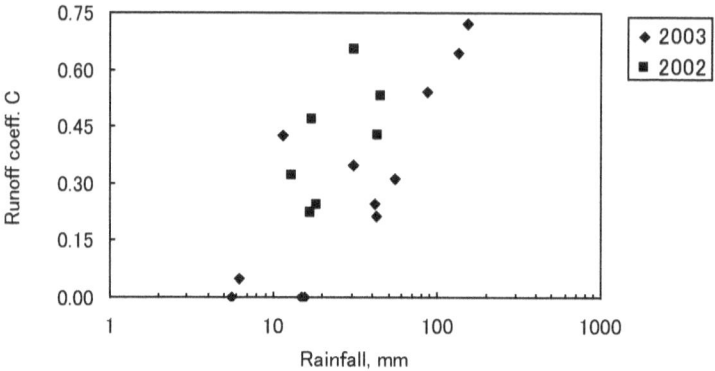

Figure 6.3 Runoff coefficient verses runoff in Khajuri main stream.

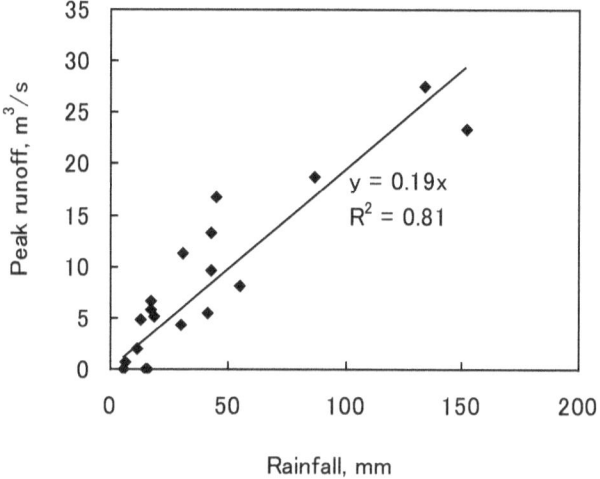

Figure 6.4 Rainfall verses peak runoff in Khajuri main stream.

Rainfall-runoff characteristics reveal that runoff generally originates from rainfalls greater than 10mm. Despite the fact that the regression equation in Fig. 6.2 indicates a good linear trend between the rainfall and runoff events; individual events show wide variation in the runoff response as evident by quite scattered plotting points. Variability in runoff response in a catchment may arise from a number of factors such as variation in rainfall intensity, antecedent moisture conditions and so on.

Infiltration characteristics

Infiltration is the process of water penetration from the ground surface into the soil matrix. Many factors influence the infiltration rate, including the condition of the soil surface and its vegetative cover, the properties of the soil such as porosity and hydraulic conductivity, and the current moisture content of the soil. Soils exhibit a great spatial variability even within relatively small areas. Moisture content of the soil varies with time. As a result of this wide temporal and spatial variation, infiltration is a very complex process that can only be defined by approximate mathematical equation (Chow et al., 1988).

When the water is pounded on the surface, the infiltration occurs at the potential infiltration rate. When the potential infiltration rate is more than the intensity of rainfall, all the rainwater infiltrates into the soil producing no surface runoff. When the infiltration rate is less, a part of rainfall begins to flow on the surface which is generally called overland flow. Therefore infiltration rate is an important soil parameter that influences the generation of surface runoff.

A number of infiltration equations are available, the basic one is given by Horton (1939). Cumulative depth of water at any time t is given by:

$$F(t) = f_c + (f_0 - f_c) e^{-kt}$$ Equation (6.2)

where, f_0 is initial infiltration rate, f_c is final (basic) infiltration rate and k is decay factor.

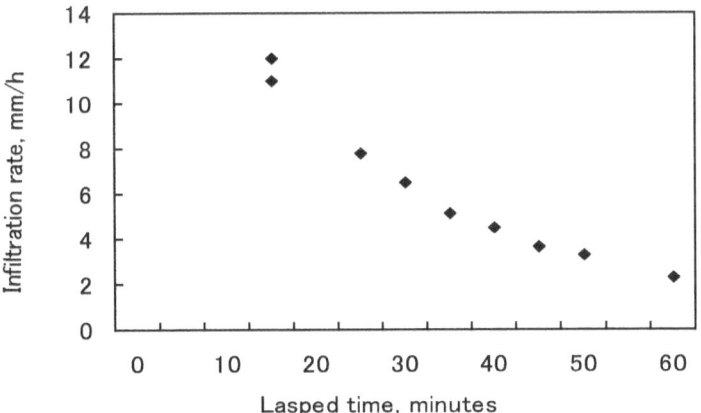

Figure 6.5 Measurement of infiltration rate at KH2 (Irasawa, 2004).

In order to examine the infiltration characteristics of the soil surface, a number of tests were undertaken in different locations in the catchment in July 2004. The tests were performed during the rainy season so that the soil surface was fairly in saturated condition. Double ring infiltrometer was used to measure the infiltration rate. Three types of land cover were selected: dense shrub land, bare slope land and partially dense forest.

Initial infiltration rate in the shrub and forest land varied from 10 to 45 mm/h. In the bare slope land, as high as 93 mm/y was recorded. Variation in the initial rate was mainly due to the embedded gravels and boulders which form pores in the soil matrix. Example of infiltration measurement is shown in Fig. 6.5.

Basic infiltration rates varied from 3 to 10 mm/h. In the area composed of silt and clay, average rate was 4mm/h whereas in the loam silt, it was about 9 mm/h (Table 6.3).

In Darcy's low, flux of subsurface flow q is defined by:

$$q = -K \, dh/dz \qquad \text{Equation (6.3)}$$

where K is the hydraulic conductivity and dh/dz is the head gradient.

Relationship between hydraulic conductivity K and infiltration rate f(t) is given by Green-Ampt equation as:

$$f(t) = K \, [\Psi \, \Delta\Theta/F(t) + 1] \qquad \text{Equation (6.4)}$$

Where, Ψ is suction head of wetting front, $\Delta\Theta$ is the change in moisture content and F(t) is the cumulative infiltration at any time t. From the equation, it is clear that f(t) is higher than K. When the soil matrix is fully saturated, then f(t) becomes equal to K.

Table 6.3 Basic infiltration rate (f_c) for different soil and land cover (Irasawa, 2004).

Location	Land cover	Soil type	Test points	Fc mm/h	Ave. fc mm/h
KH2	Dense shrubs	Silty clay	1	3	
	Dense shrubs		2	4	4.0
	Dense shrubs		3	5	
MH1	Bare	Laterite silt loam	1	4	
	Bare		2	10	6.3
	Bare		3	5	
MH1	Partially dense forest	Loamy silt	1	8	
	forest		2	10	9.3
	forest		3	10	

Infiltration parameters for the Green-Ampt equations are given in Table 6.4.

Table 6.4 Hydraulic conductivity for various soil classes (Rawls et al., 1983 cited in Chow et al., 1988).

Soil class	Hydraulic conductivity K mm/h
Sand	117
Loamy sand	30
Sandy loam	11
Loam	3.4
Silt loam	6.5
Sandy clay loam	1.5
Sandy clay	1.0
Silty clay	0.5
Clay	0.3

In the view of the fact that infiltration rate should be higher than K, the measured rates seem plausible if compared to the K-values in Table 6.4.

The infiltration rates however seem much smaller when compared to the rainfall intensities in the study area where generally rainfall events greater than 10mm have maximum intensity greater than 20 mm/h. The forest cover seems not contributing significantly to increase the infiltration rate mainly due to the soil crust, which is made up of fine materials. This type of soil surface with lower infiltration rates may, to some extent, have contribution to the generation of quick flow in the study area. However, due to the limited experiment sites, caution must be taken when interpreting these results in case of other catchments.

Estimation of peak discharge

Estimation of peak discharge is an important task for designing protection works such as embankment. So, an attempt was made to apply Rational formula for estimating peak flows, which is a simple and most widely used method in unguaged small catchments. The method is based on the principle that maximum rate of runoff from the catchment appears when the entire area contributes at the basin outlet. The runoff gradually increases from zero to peak when rainfall duration reaches the time of concentration and it becomes constant onwards. The equation is given by:

Q=CIA/360 　　　　　　　　　　　　　　　　Equation (6.5)

where, Q is peak flow (m^3/s), C is runoff coefficient, I is maximum rainfall intensity (mm/h) for the duration equivalent to time of concentration and A is the drainage area (ha).

There are numerous equations available to compute time of concentration T. Most of them use the empirical equations relating the catchment variables like length of channel and catchment area. Here, the method developed by Rziha is used:

Rziha equation: T= L/W and W=20$(h/L)^{0.6}$ 　　Equation (6.6)

where, T: time of concentration (sec), L: Length of catchment along the longest channel (m), h: elevation difference between the ridge line and point of measurement (m)

　　Taking L = 3.68 km and h= 140m, for Khajuri stream and L = 2.63 km and h= 140m, S= 2.5 % for Mushahar stream, the value of T comes to be 22 minutes and 15 minutes for the two streams respectively. The estimated time of concentration seems reasonable because the time to maximum peak flow was observed to vary between 20 to 40 minutes.

　　Since the durations of maximum intensities recorded were generally above these values, maximum intensities were taken to calculate the peak flow using equation 1. The runoff coefficient (C) is the least precise variable of the rational method, and proper selection of this value requires a sound judgment and experience (Chow, 1964). So, various trials were made for C in a bid to match the estimated and measured values. However, none of the selected values could yield satisfactory result (Fig. 6.6).

　　The result obviously reveals the shortcomings of the Rational method. In this method, runoff coefficient is assumed to be the same for all storms which means that the losses are constant for all storm sizes. However, the losses vary according to the rainfall patterns. The proportion of the rainfall that will reach the outlet point depends upon various factors such as percent imperviousness, ground slope, degree of porosity and so on. Runoff generation is closely related to the antecedent moisture condition, which again depends on the time period and frequency of the rainfall events.

Measurement of suspended sediment

Measurement of event runoff and suspended sediment was undertaken in the rainy seasons of 2002 and 2003. Concentration of suspended sediment was measured taking a number of flood samples during different stages of flood event. For each flood event, at least two samples were taken during the rising period, one during peak and two during the recession period. Concentration of suspended sediment varies significantly event-by-event and even in the same event. The relationship between discharge and sediment concentration is called sediment-rating curve. The curve for the Khajuri stream is shown in Fig. 6.7 which indicates a very widely scattered plot of concentration data. Most of the data points are above 10 gm/lit.

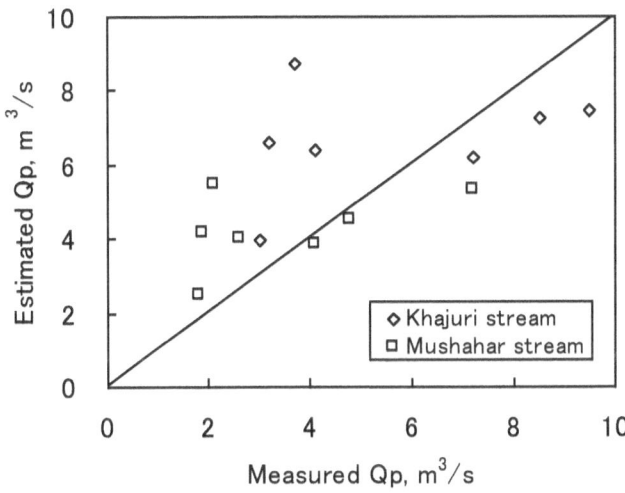

Figure 6.6 Measured verses estimated peak discharge (C= 0.27).

Fig. 6.8 shows the plot between suspended sediment yield from rainfall events. It is important to note that the sediment-rating curves are for the Khajuri and Mushahar streams together as computation has been done taking the runoff data downstream of the confluence point. Event wise sediment yield was estimated by multiplying the sediment concentration by runoff volume computed from flood hydrographs. The figure shows a generally increasing trend of sediment yield with the rainfall size.

Virtually there is no data available on the measurement of sediment in the Siwalik Hill region, which does not warrant the comparison of the measured sediment data. In the Middle Hills, Merz (2004) reports the sediment concentration in discharge from different plots and watersheds, which are basically under agriculture. The concentration was found to decrease with the increase in catchment size clearly indicating the scale dependence of sediment mobilization process.

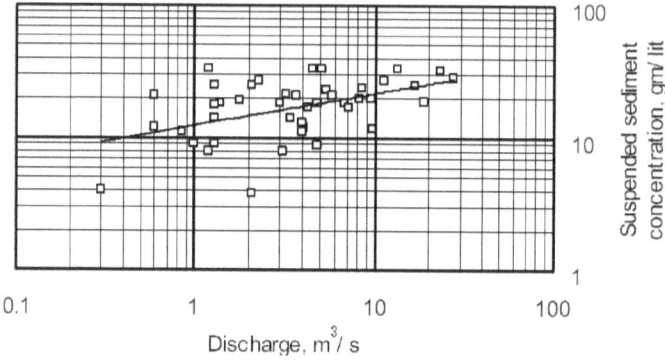

Figure 6.7 Suspended sediment concentration in Khajuri stream.

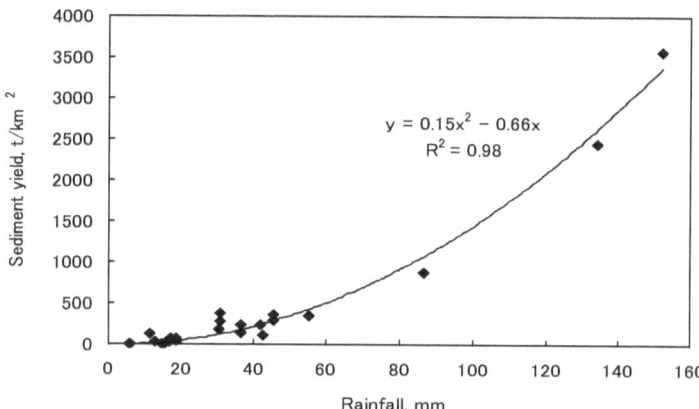

Figure 6.8 Suspended sediment yield from different rainfall events in the Khajuri main stream.

It is apparent that the number of sediment data is not sufficient to draw a concrete result on the sediment mobilization; however it can provide basic understanding on the order of magnitude of sediment losses from the catchment. Based on this, annual sediment yield has been estimated which is then compared to the sediment production rate from the catchment estimated with the help of field measurements.

Summary

Following points can be summarised:

- The streams are characterized by small basin area and steep slope gradient resulting in a high velocity of discharge flow.
- The floods are of temporary types with very short flow duration, popularly known as flash floods. Their flow pattern is generally not predictable.
- Peak flow occurs within a very short time once the discharge starts. It generally occurs within 30 to 40 minutes in the study streams. The reason of such quick generation of the peak flow is assumed to be the high drainage network and steep topography in the upstream reach of the catchment.
- Rainfall-runoff analysis indicates that there was no perfect linear trend in between them. Runoff coefficient varies from almost zero to as high as 0.72 depending on the rainfall characteristics.
- Basic infiltration rates vary from 4 to 9 mm/h for various soil and land covers. Relatively these rates seem much lower compared to the rainfall intensity in the study area, which may be a reason to explain the quick generation of peak flow.
- Rational method to estimate the peak flow could not yield satisfactory results. Main reason could be that runoff coefficient can not be a constant as assumed in the method.
- Measurement of suspended sediment shows that majority of events had concentration more than 10 g/lit. The concentration generally increases with the increase in rainfall and runoff.

7
Identification of sediment sources

Background
Identification of type and distribution of sediment sources is an important step for understanding the sediment mobilization process in a given catchment. It is also important for the estimation of sediment production in the catchment scale.

Key sources of sediment
The sources of sediment can broadly be divided into two types: non-channel and channel. The relative contributions of the sources tend to vary with distance downstream. In general, the upper parts of catchments, with their steeper slopes and stronger coupling between hillslopes and channels, supply sediment largely from non-channel sources (Knighton, 1998). Further downstream where slopes are gentler and floodplains wider; the potential for temporary storage of eroded material increases and the contribution from channel erosion becomes more important.

Non-channel sediment sources are associated with various erosion processes occurring in the hillslopes in the form of surface, rill, gully and landslide erosion. Erosion processes in the channel include stream bank and channel bed erosion. These processes operate at different scales with wide temporal and spatial variation. Process dominance generally varies longitudinally in a stream channel (Schumm, 1977).

Distribution of sediment sources

In order to categorize the distribution pattern of principal sediment sources in the study catchment, geomorphological classification can be taken as a basis. It has been found that type and distribution of the sediment sources are closely related to the geomorphological classification, which implies that there is dominance of specific erosion process in specific geomorphologic unit (Fig. 7.1).

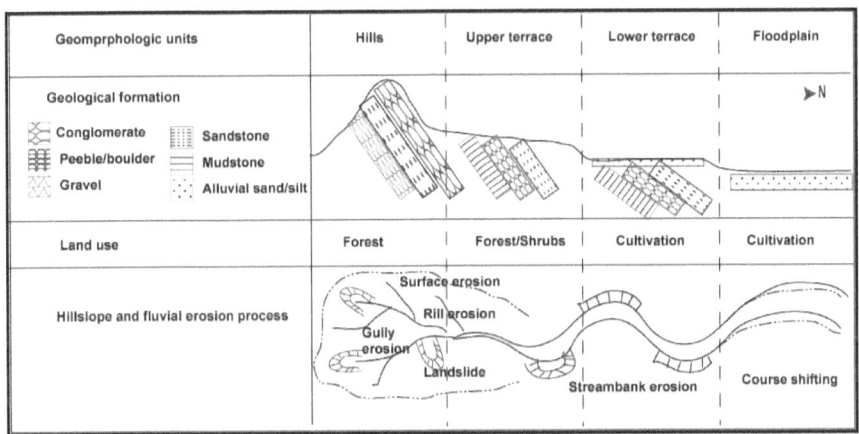

Figure 7.1 Sketch depicting erosion processes in different geomorphological segments.

As presented in Chapter 2, the study catchment can be divided into four geomorphologic segments: hills, upper terrace, lower terrace and floodplain. The hills which are composed of coarser sedimentary rocks, are characterized by frequent instabilities despite thick vegetation cover. Many rills, gullies and landslides dissect the hills making it a major source of sediment.

The ridge and watershed divides are sharp and narrow, which indicates the prevalence of rapid surface erosion. Most slopes are formed under the combined rather than the unique action of mass movement and running water. In the hills, since the streambanks are composed mostly of bed-rocks, they do not generally show the active cutbanks. In the terrace reach, however, channel side scouring is predominant forming tall cutslope banks. The upper terrace generally consists of higher cut banks than the lower one does. Tall cutbanks are distributed on both sides of the streams. In the floodplain, there is no clearly defined streambank, which is often characterized by change in stream morphology. Slope instabilities in the catchment were mapped by aerial photo interpretation, which clearly indicates that landslides and large gullies are mostly clustered in the hills (Fig. 7.2).

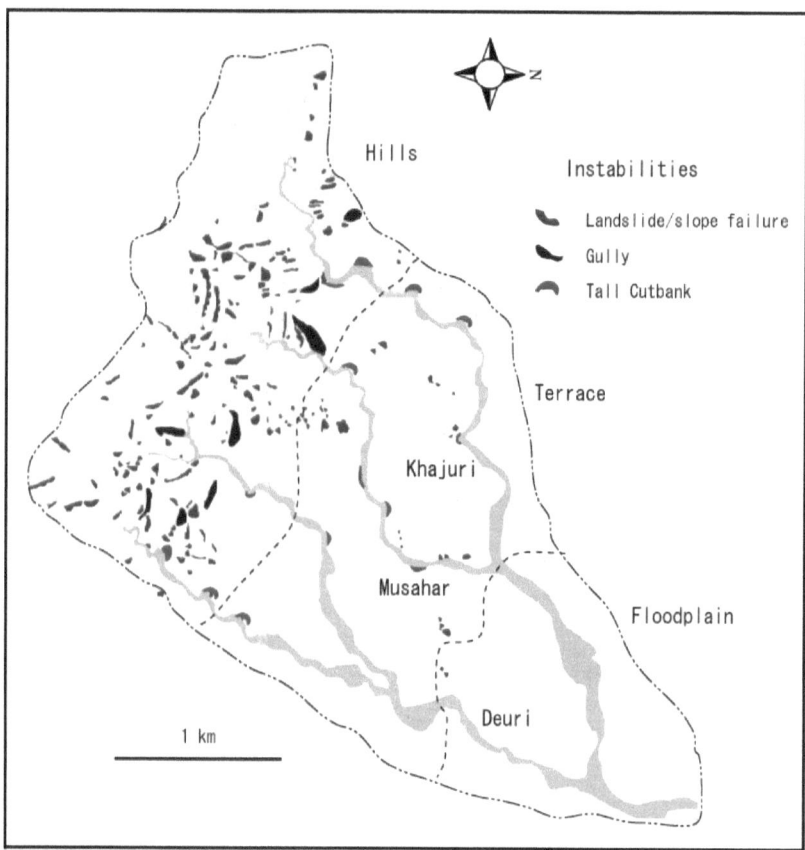

Figure 7.2 Distribution map of sediment sources in Khajuri and Deudi catchment (Interpreted from 1992 aerial photo, 1:50000 scale with field verification).

This type of close relationship between the distribution of sediment sources and geomorphological classification is helpful for identifying key sediment sources and establishing basis for studying many geomorphological processes in the regional scale.

Summary

Following points can be summarised:

- The area can broadly be classified into four geomorphological segments: hills, upper terrace, lower terrace and floodplain, as discussed in Chapter 2.

- Distribution of the instabilities (or sediment sources) largely depends on the geomorphological classification. Landslides and gullies occur mostly in the hills, while active cutbanks occurs in the terrace reach. Low-height cut-banks form mostly in the lower terrace.

8
Hillslope erosion processes

Background

As described in the research framework, four types of processes are identified: surface erosion, rill erosion, gully erosion and landslide slope erosion. Each process has been described on the basis of intense field observation and monitoring. Particular emphasis has been given to estimate the long-term average rate of these erosion processes within the catchment. Key driving factors are also described to help understand these processes at a catchment scale.

Surface erosion

This section details the process of surface erosion on the hillslopes of the Siwalik Hills. The processes are closely related to topography, soil and land cover. It gives the information on the erosion plots set up to monitor erosion processes, monitoring methods and interprets the data considering the key controlling factors.

Layout and design of erosion plots

Land use can significantly affect the magnitude of sediment yield through its influence on the degree of protection afforded by vegetation cover, physical characteristics of soil and potential for surface runoff generation (Knighton, 1998). Hence, four erosion plots were set up covering different land covers (Fig. 8.1, Table 8.1). The location of the plots was chosen considering that they can represent other areas with similar land cover within the catchment. The plots were established on the ridge slopes (upper slope of the hills) where erosion mostly occurs by overland flow. Erosion pins were fixed at 3 to 5m intervals forming square or rectangular grid network (Fig. 8.2). The slope gradient of the erosion plots varies from 10 to 35 degrees. Slope length varies from 9 to 38 m. Plots on the forest land generally consist of black soil containing mostly silt and fine sand. Plot on the bare land however consists of red soil composed of silt and fine sand.

Figure 8.1 Erosion plots.

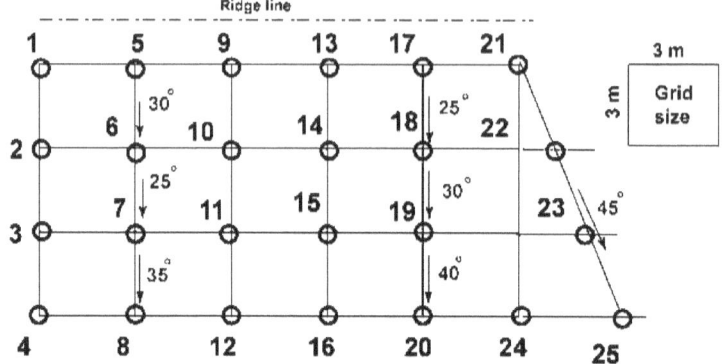

Figure 8.2 An example of measurement grid at KH1 with a grid 3m x 3m.

Table 8.1 Characteristics of erosion plots.

Site	Land use	Av. slope (degree)	Size (m x m)	Slope length (m)	Soil type	No. of erosion pins
KH1	Partially dense forest	30	15 x 9	9	Black soil, Silt and fine sand	19
KH2	Shrub and bush	10	38 x 10	38	Silt and fine sand	11
KH3	Partially dense forest with shrub	20	19 x 11	19	Black soil, Silt and fine sand	15
MH1	Bare	35	24 x 12	24	Red soil, Silt and fine sand	21

Rate and process of erosion

Annual average rates of erosion from each plot are computed in Table 8.2, while erosion rates from individual peg are shown in Fig. 8.3. Table 8.2 shows that the average annual rate from the plot in the forest cover KH1 is 1 ± 0.6 mm. Similarly, it is 0.9 ± 1.8 and 1.2 ± 0.7 mm from the plots on shrub land and partially dense forest, KH2 and KH3 respectively. Thus from the forest and shrubland, no significant difference was found in the erosion rates. In contrary to this, bare land is found to be suffered by significantly higher erosion rate which is 7.0 ± 3.8 mm per year. The error limits are taken as the 95% confidence limits. It is important to note these rates have been computed considering the erosion only during the rainy season (June-September). It is reasonable because majority of erosion takes place during this period.

Table 8.2 Average erosion rates.

Site	N	Erosion rate, mm/y					Monitoring period
		Mean	S.D.	Max	Min	*SE	
KH1	19	**1.0**	1.2	3.3	-1.2	0.6	May 02-Aug 04
KH2	11	**0.9**	3.1	5.0	-4.0	1.8	May 03-Aug 04
KH3	15	**1.2**	1.3	4.0	-1.5	0.7	May 03-Aug 04
MH1	21	**7.0**	9.0	33.3	-1.7	3.8	May 02-Aug 04

* Standard error limits are the 95 % confidence interval limits.

(-) values indicate deposition.

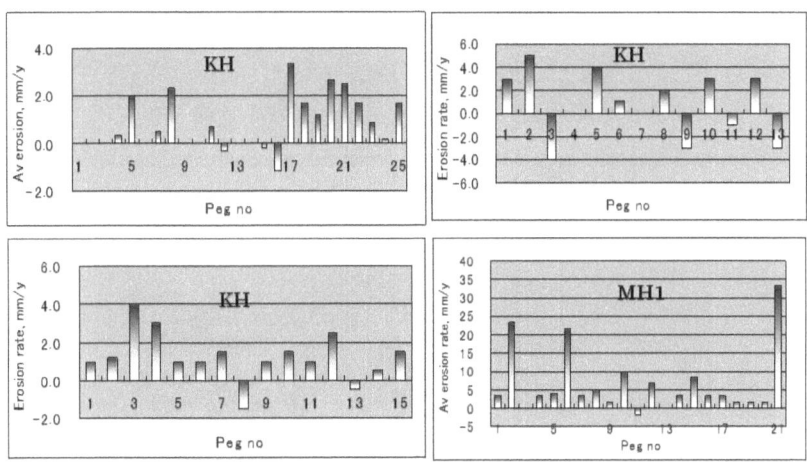

Figure 8.3 Average erosion rates from individual peg.

Fig. 8.3 reveals a wide spatial variation in erosion rates at a plot scale. Monitoring data indicates that most of the erosion as expected took place during the rainy season. There were many pins with negative erosion indicating soil deposition mostly in the forest and shrub land. This was apparently due to the fact that surface litters intercept runoff flow and sediment to some extent. The inconsistency in the soil loss rates indicated by individual erosion pins in a plot can partially be explained by the variation in micro-topography. In KH1, for example, erosion rates of pegs from 17 to 25 are higher than from 1 to 17. This difference may be explained by the fact that slope gradient (average 45 degree) belonging to peg 17-25 is higher than the pegs 1-17 (average 30 degree) (Fig. 8.2).

There may also be other inconsistencies in a plot because of difference in plant cover and litter distribution, however, inconsistency in erosion rates could not be explained by such differences. At MH1, higher erosion rates are found at the pins located along small rills (e.g. pegs 2, 6 and 21, Fig 8.2). However no clear linkage was noticed between the variation in erosion rates and topographical difference for the plots KH2 and KH3.

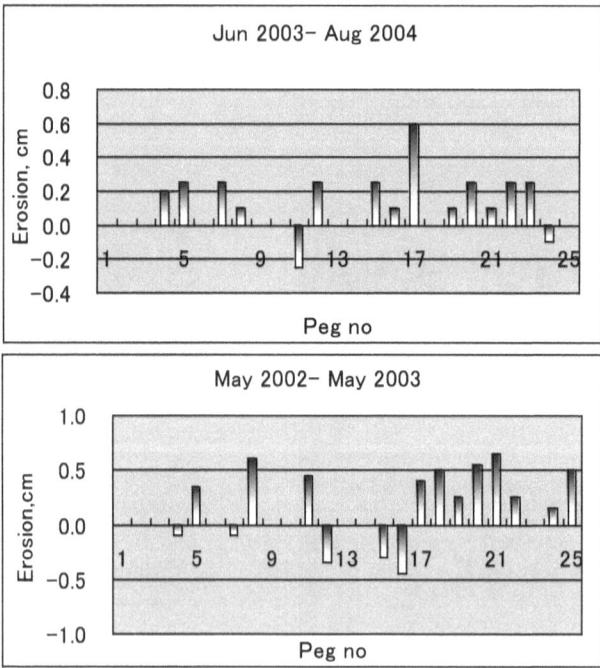

Figure 8.4 Erosion amount from each peg in KH1 during the two monitoring periods. (Note that pegs 1,2,3,6,10,13,14 were lost during monitoring).

On the other hand, wide temporal variation in erosion rate was observed during the monitoring period. At KH1, for example, erosion amounts during the two-year monitoring were remarkably different from each erosion peg (Fig. 8.4).

The first monitoring period (May 2002 to May 2003) includes one rainfall season and had average erosion amount of 1.8 mm whereas the second one (Jun 2003 to Aug 2003) contains two rainfall seasons with average erosion amount of 1.4 mm. There were some pegs that demonstrated both erosion-deposition phases. When analyzing rainfall pattern during the monitoring period, no clear linnk was observed between rainfall size and erosion amount of each peg. Similar trend was observed for other erosion plots too.

Controlling factors

There are host of controlling factors and modes of occurrence of surface erosion from hillslope. Soil character, slope gradient and ground cover are the most dominant factors. In the study catchment, soil type is almost similar in the forest and shrub lands. Most of the area is dominated by silt and fine sand (Table 8.1). In contrast, bare lands have undergone intensive weathering resulting in the formation of red soil. Due to lack of organic materials, red soils are particularly sensitive to degradation (Shah et al., 2000). It is generally accepted that erosion rate increases with the increase in slope gradient. However, having the similar slope gradients, the bareland plot (MH1) loses on average about seven times as much soil as does the forested plot (KH1). Although this difference in erosion rate may arise partly from the difference in soil formation, effect of ground cover is more apparent.

Ground vegetation has an important role in reducing soil loss. It acts as a protective layer between the atmosphere and the soil. Vegetation has mainly two hydraulic functions: firstly, it intercepts the direct rainfall which retards the erosivity of raindrops; secondly, it retards the velocity of surface runoff which promotes infiltration. In addition, well-developed root systems can strengthen the soil structure. Ground cover especially low height trees and shrubs play important role in retarding soil loss (Morgan, 1995). The plots in the forest and shrub lands did not have significant difference in erosion rates mainly due to the fact that both contain low height plants. In addition, both types of plots were covered by enough fallen leaves and litters. Siwalik Hills are mostly covered by Sal trees (Shorea robusta) which are of deciduous type producing significant litters on the hillslopes. It may be pointed out that water run-off, infiltration capacity and soil erosion in the Siwalik region are highly dependent on the vegetal cover and leaf litter (Singh, 2001).

Although relatively easy and simple to apply, erosion pin method is less accurate than other methods such as collector trench in assessing soil loss from a hillslope. Moeyersons (1990) describes advantages and limitations of this method. First, main important source of error is, inserting of erosion pin makes the soil surface loose to some extent which can produce excess erosion than reality. Second, it can be disturbed by ant activities. Third, erosion nails can be lost due to disturbances by human. In order to minimize the first source of error, thin iron nails (8mm diameter) were used. Proper care was taken while inserting the pins into the surface. They were inserted with uniform impacts of hammering in order to minimize the soil disturbance. For lessening second error, the plot locations were fixed away from the area of ant activities. A number of pins were missing during the monitoring period by human and animal activities; however they were replaced immediately in order to establish data over a more representative period.

Accuracy of data largely depends on the measurement method since erosion figures especially in the forested land are rather small. Therefore measurement was carried out using measuring hand tape with precision of 1mm. This precision is practically suitable since it is virtually impossible to measure erosion less than 1mm especially in the forested slope which is generally rich in ground litter. In order to ascertain the erosion level during repeated measurements, the component of the pins exposed above ground surface was enamel painted. Although the data could have other unavoidable errors, it is urged that they fairly represent the erosion processes with reasonable accuracy.

Summary

Following points can be summarised :

- Average annual rate of soil erosion: 1.0 ± 0.6 mm for dense forest, 0.9 ± 1.8 mm for partially dense forest, 1.2 ± 0.7 mm for shrubs and bushes, and 7.0 ± 3.8 mm for bare land.

- There were wide temporal and spatial variations in the erosion rates.

- No clear proportional relationship was observed between rainfall size and erosion amount of each peg.

- Effect of ground vegetation is more conspicuous than the effect of slope gradient. Low height plants and litter covers are the important controls of surface erosion.

Rill erosion

This section describes the rill erosion processes with some field measurements. Types, forms and rates are described briefly based on the field observations.

Fundamentals of rill erosion

Rills are small, closely spaced channels with cross-sectional dimensions of a few centimeters to a few tens of centimeters (Selby, 1993), which results from the uneven erosion of soil by running water. Rills are comparatively smaller than gullies but there is no clear demarcation in sizes between them. Gullies are relatively deep and have clear headcut in comparison to the rills.

It is widely accepted that rills are initiated at a critical distance downslope where overland flow becomes channeled. The separation of shallow overland flow into small channels produces secondary flowpaths. Where these converge, the increase in discharge intensifies the particle movement initiating channel scouring. Morgan (1986) mentions that change from overland flow to rill flow passes through four stages: unconcentrated channel flow, overland flow with concentrated flow paths, micro-channels without headcuts and micro-channels with headcuts. The process of rill erosion in terms of rill initiation and headcut development has been described in many studies (e.g. Govers, 1985; Slattery and Bryan, 1992; **Abrahams et al., 1996; Nearing et al, 1998**).

On one hand, rills are usually discontinuous which may have no connection to a stream-channel system and they are often obliterated between some storms. On the other hand, rills may enlarge in the form of gully which can form the heads of natural drainage systems. Both types of rills can be observed in the study area.

Forms and processes

Three types of rills were observed in the field, which are described here briefly.

(1) Vertical rill on the steep and tall cutslopes.

Whether it is on the slopes of gully heads or stream banks, rill erosion can be observed mainly in the formation of fine grained soil usually clay and silt with interbedding of coarse sand and gravel. Numerous rills develop almost in parallel beneath the head crown of such slopes, which then coalesce to form micro-channels (Fig. 8.5). These are formed largely by the impacts of raindrops

(2) Inclined rills on the slopes of landslides/slope failures

The slopes formed by landslide scarps demonstrate active rills which are generally larger in size than the vertical rills described in (1). Such rills are often in the size of 30 to 50 cm in width and 10 to 20 cm in depth. Raindrops as well as surface runoff are the triggering factors of such rills.

(3) Inclined rills developed on the hillslopes

These are the rills that develop on the hillslope surfaces essentially as a result of concentration of surface runoff flow. Obviously the size of rills depends on the runoff contributing area. Therefore, rills initiated from the hillslope crown gradually increase in size downslope, which ultimately develop into small channel usually up to one meter in depth before merging into the main stream.

Figure 8.5 A steep head of a gully featuring rill erosion on its mid-slope. The formation is silty clay interbeded with gravel and boulder (Left). Vertical cutbanks made of silty clay with interbedding of sand. Note that the rills combined together to form a hollow following piping failure (Right).

Measurement of rill erosion

The third type of rill was monitored in the study area. For this, three rills were established based on land cover- on bare land, shrub land and forest. For rill on bare land, the same bare slope for surface erosion (MH1) was selected while other two were in the vicinity of the erosion plots (KH1 and KH2). The characteristics of the selected rills are shown in Table 8.3. The location of each monitoring site is shown in Fig. 3.2 (Chapter 3).

It is noteworthy to mention that it was difficult to ascertain the exact width, as the rills were mild concave in shape. Hence, only bed erosion was monitored instead of changes in sectional area. The forested rill R3 which was established one year later than the other two had become full of litters deposited during the rainy season of 2004. As a result, measurements were obstructed. Hence, no quantitative rates could be obtained from this site. However, it does indicate that rill erosion from the litter rich forest cover is insignificant compared to that from bare land.

Table 8.3 Characteristics of some selected rills and erosion rates.

Rill	Land use	Slope gradient degree	Runoff contributing area, m^2	Av. rill size W cm x D cm	Surveyed length, m
R1	bare	35	50	20 x 20	16
R2	shrubs	22	400	40 x100	15
R3	forest	10	80	15 x 10	10

Site	N	Erosion rate, mm/y				
		Mean	S.D.	Max	Min	*SE
R1	9	14.0	10.4	30.0	2.7	6.8
R2	6	10.0	6.7	12.5	-7.5	5.3

* Standard error limits are the 95 % confidence interval limits.
(-) values indicate deposition.

Erosion rates are shown in Fig. 8.6, which indicates that average rates of rill (bed) erosion at R1 varied from 3 to 31 mm/y, with an average of 14 ± 6.8 mm/y. If compared this value to the surface erosion from the same plot (MH1), rill erosion is around double the rate of surface erosion (7.0 ± 3.8 mm/y). Like in the surface erosion, there were wide variation in erosion rates both in space and time. The figure also reveals that average erosion rate at R2 was 10 ± 5.3 mm/y. At R2, many negative erosion (deposition) points were recorded. Apparently it is due to the fact that erosion pins trapped sediment with fallen leaves and litters. However, sediment deposition was also observed at the locations other than the pins. It was mainly prevalent in the narrower section of the rills, especially in the upstream as shown in the Fig. 8.6.

Even though rill erosion is very important process in the wake of degrading land cover in Siwalik Hills, detailed process studies could not be undertaken in this study. Nor they could be incorporated in the sediment yield estimation from the catchment because of difficulty in estimating the rill density in a hillslope. However, since the erosion pins were located randomly on the slope surface, the surface (inter-rill) erosion incorporates, to some extent, micro-rills formed due to uneven micro-scale topography. It can compensate for rill erosion to some extent.

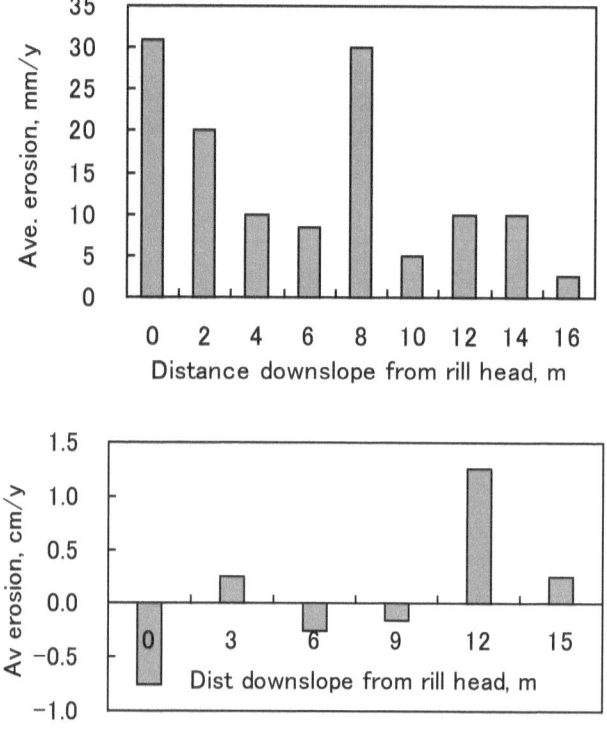

Figure 8.6 Erosion rates from rills, R1 (above), R2 (below).

Summary on rill erosion

Following points can be summarised:
- Rills vary widely in space in terms of shape and size.
- Vertical rill erosion which occurs mainly by raindrop impacts is witnessed in the steep headwall of gullies and sidebanks composed of fine-grained silt and clay. In the headscarp slopes formed by landslides/slope failures, inclined rills are predominant which occur mainly by combination of raindrops and surface flow. In the natural slopes, rill erosion occurs mostly by overland flow

- Rate of rill erosion also varies widely in space and time. Average rates of rill (bed) erosion at R1 varied from 3 to 31 mm/y, with average 14 ± 6.8 mm/y, which is twice larger than the rate of surface erosion on the same slope.
- Amount of rill erosion is closely related to runoff contributing area.
- Erosion-deposition phases were observed in the rills formed in gentle slopes of shrubs and forest, indicating that fallen leaves and litters have significant influence on rill erosion.

Gully erosion

This section introduces the key geomorphological features of gullies with a comprehensive examination of the erosion processes. Special consideration is given to the development processes of gully head because of the fact that much more sediment is generated from the head area than from the gully channel. Rate of erosion is estimated based on comparison of aerial photographs as well as field monitoring and discussion is made on the governing factors.

How are gullies formed?

Gullies are relatively deep, unstable, eroding channels that form at the head, side or floor of valleys where no well-defined channel previously existed (Schumm et al., 1984). Compared with stable river channels, which have a relatively smooth, concave-upwards long profile, gullies are characterized by a headcut and various steps or knick-points along their course (Morgan, 1986). Gullies also have relatively greater depth and smaller width than stable channels, carry larger sediment loads and display very erratic behaviour so that relationships between sediment discharge and runoff are frequently poor (Heede, 1975).

Gully expansion process occurs mainly by gully-head erosion. As a result of this process, gully move upslope, releasing sediment to the channels and exposing new channel walls to erosion. Hence, depending upon the rate of retreat, gully erosion may represent an important sediment source in a range of environments (Poesen et al., 2003) and also act as sensitive indicator of environmental change (Oostwoud Wijdenes and Bryan, 2001). Because they are very rapidly developed erosional forms, gullies are usually not regarded as features of normal erosion, but the result of the changes in the environment such as faulting, burning of vegetation, overgrazing, climatic change, extreme storms or any other cause of break in vegetation which will bare the soil (Selby, 1993). Gullies are almost always associated with accelerated erosion and therefore with landscape instability. For that reason, the rate of gully growth is one of the most important indicators used for assessing the area destroyed by gully erosion and evaluating many kinds of damage it causes. Advancing gully heads cause damage to roads and structures, and may result in the loss of forest, agricultural, residential and recreational land (Seginer, 1966).

Types and process of gully erosion

Gully erosion generally starts for one of the two reasons: either there is an increase in the amount of flood runoff, or the flood runoff remains the same but the capacity of water courses to carry the flood waters is reduced. The most common causes of increases in runoff or deterioration in channel stability are the changes in vegetation cover- especially removal of trees, increase in the proportion of arable land in the catchments, excessive removal of vegetation or over-grazing, or a climatic change with accompanying variations in rainfall periodicity and intensity (Silby, 1993).

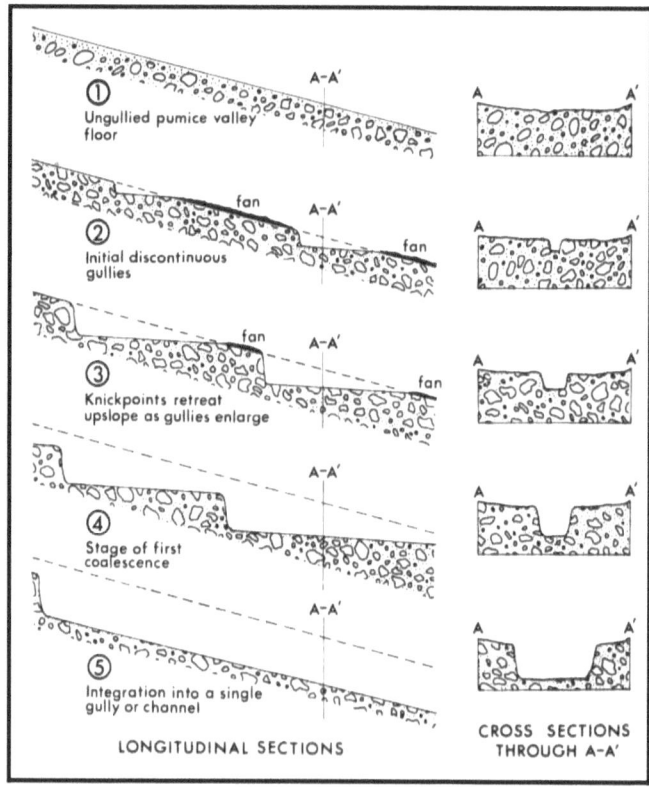

Figure 8.7 Stages of development of discontinuous gullies (Leopold et al., 1964, modified by Silby, 1993).

Gullies can be classified into two types broadly, although compound types are also common: (i) continuous gullies and (ii) discontinuous gullies. A continuous gully which forms by enlargement of a rill may have no headscarp because it forms in the non-cohesive materials without a resistance capping. (Selby, 1993). Such gullies can also develop within the scars and deposits of large mass movement features (Morgan, 1988).

Discontinuous gullies, on the other hand, may develop at several points where vegetation cover is broken. If the gully originates at a scarp or steep face and the gully is cutting into cohesive materials, they maintain to develop the headscarps. In contrary, the headscarp is not maintained in the non-cohesive materials.

Leopold et al. (1964) postulated different stages of development of discontinuous gullies (Fig. 8.7). In the first stage, small depressions or knick points form on a hillside or a valley floor as a result of localized weakening of the vegetation cover (such as by cattle tracks, pot-holes, grazing etc.). Water concentrates into these depressions and enlarges them until several depressions coalesce, and an incipient channel is formed. Ultimately a single gully is developed by upslope retreat of the knick points. Another form of gully development occurs through the process of sub-surface flow called "piping". It is predominant in the soil rich in clay content which has strong swelling and shrinkage properties. The soil surface cracks are developed as a result of desiccation. During rainfalls, water infiltrates rapidly down the cracks and supersaturates a relatively permeable horizon in the sub-soil. Lateral seepage moves the soil particles, which can initiate and enlarge a pipe in the form of gully.

Previous studies

Occurrence of gullies varies widely as regards to type, form, impact and development. Complexities in the gully erosion process have been outlined in voluminous literatures. Most of the studies have been concentrated in the arid and semi-arid environments where rainfall amount is too small as compared to the tropical and sub-tropical environments. Though gully erosion is a very common problem, not many studies are undertaken in the Asian monsoon region. Some of the studies undertaken in other parts of the world are summarised below.

Gully expansion, usually measured as an increase in area and/or length over time, is commonly evaluated from successive aerial photographs or maps (Beer and Johnson, 1963; Thompson, 1964; Seginer, 1966; Stocking, 1980).

Gully expansion process is often linked to a variety of factors including drainage basin area, gully dimension parameters, indices of surface runoff and precipitation, antecedent precipitation, soil moisture, and indices of piping (Beer and Johnson, 1963; Thompson, 1964; Seginer, 1966; Piest et al., 1975; Stocking, 1980). Burkard and Kostaschuk (1997) listed the parameters used to examine gully growth by many researchers as shown in Box 8.1.

Vandekerckhove et al. (2000) investigated the morphology of actively eroding bank gullies and relationship between the controlling parameters in the Mediterranean environments. They found that erosion processes shaping the gullies are strongly related to the soil material characteristics. They observed piping in the soil layers with higher silt content, lower sand content and a higher electrical conductivity. They also observed a negative relationship between the local slope at the gully head and the present drainage basin area.

Martin- Penela (1994) studied the evolution of gully systems and badland development in Almanzora river basin in the Mediterranean. He found a strong litho-structural control on the creation and evolution of piping and gullying processes. The gullies developed in two different topographic contexts: one on the very steep slopes and another on the sub horizontal surfaces of abandoned agricultural land.

Oliveira (1990) found that gully development in the Bananal area of Brazil was strongly depended on the geometry of the hillslope which is defined by the slope, angle, length, height and area. Gully erosion was concentrated on the concave slopes where three types of gully development were distinguished: (1) gullies connected to the drainage network located on footslopes where the dominant process was seepage erosion, (2) gullies not connected to the drainage network located on the higher slopes where dominant process was concentrated overland flow and (3) a combination of these two types, connecting upper and lower slopes.

Ebisemiju and Ekiti (1989) undertaken a morphometric approach to analyze the erosion processes in the valley-side gullies formed in the laterite terrain of the northern Nigeria. They examined the inter-relationships and variations statistically in the morphological properties and causative factors. They found the cross-sectional and longitudinal variables to be the prime orthogonal dimensions of gully systems. Gully cross-sections variables were found more strongly inter-dependent than the parameters describing gully longitudinal profile.

Vandekerckhove et al. (2000) studied about the topographical thresholds for gully initiation and sedimentation in the intensive cultivated fields and rangelands in the Mediterranean area of Europe. They found a negative power relationship between the local slope of soil surface and drainage basin area at the gully initiation points. The channel initiation was found to be influenced by overland and sub-surface flow.

> **Box 8.1 Parameters used to examine gully growth** (Compiled by Burkard and Kostaschuk, 1997 reviewing various studies).
>
> - Basin area
> - Basin length
> - Basin height
> - Channel slope
> - Total precipitation
> - Antecedent precipitation
> - Runoff
> - Headcut advance rate
> - Areal growth rate of gully
> - Clay content
> - Soil moisture
> - Soil shear strength
> - Soil bulk density
> - Ground water level changes
> - Season
> - Vegetation cover
> - Index of piping
> - Rainfall interception
> - Population density
> - Agricultural use

Gully head morphology and its relation to the development processes were examined by Oostwoud Wijdenes et al. (1999) in the abandoned fields located in the semi-arid environment of southeast Spain. Four categories of head morphology was defined: gradual, transitional, rill-abrupt and abrupt. They found that abrupt headcuts were always formed from secondary headcuts in the channel which migrated upstream. Gradual types of headcuts were controlled by fluvial processes, and abrupt headcuts by the combination of fluvial and mass-wasting processes. The rill-abrupt types of headcuts were still actively retreating.

Oostwoud Wijdenes and Bryan (2001) examined gully head erosion processes and estimated sediment budget in a semi-arid valley floor in Kenya. They observed gully retreat rates ranging from 0 to 15 m/y. In the annual timescale, they found that rainfall amount was a good indicator of gully head retreat while at storm-event timescale, rainfall distribution was more important. They established a relationship between retreat amount and rainfall properties- amount and number of dry days. Different components of erosion budget were found to be closely related to the number of dry days between storms.

Crouch and Blong (1989) developed a method for classification of gully sidewalls and dominant processes from a study in eastern Australia. They observed that active sections of many gullies were dominated by particular side forms. Fluting was the most prominent both in frequency of occurrence and in the proportion of sidewalls affected. In some gullies, wall failure and gully side undercutting were more important processes.

In the recent years, some studies are conducted on the modeling of gully erosion. Nachtergaele et al. (2001) evaluated the Ephemeral Gully Erosion Model (EGEM) in predicting the rate of soil erosion from ephemeral gullies in the Mediterranean environment and found a good agreement in measured and predicted rates when the gully length is given as input to the model. However, the agreement was weak when the estimated and measured volumes of soil loss were divided by the length. So they concluded that the model was not fit in the Mediterranean condition.

The use of digital elevation model (DEM) in combination with sequential aerial photographs is increasingly in use in recent times in order to estimate the rate and pattern of gully development. A recent study was done by Martínez- Casasnovas et al. (2003) on the rate of erosion from large gullies in the Mediterranean area of Spain using multi-date digital elevation model (DEM). The study shows that the gully erosion serves significantly to the total soil loss. A sediment production rate of 846±40 t/ha/y was computed for a gully area of 0.10 km^2.

Derose et al. (1998) used sequential digital elevation models (DEMs) to estimate sediment production from gully erosion in the Mangatu forest of New Zealand. In the period from 1939 to 1958, the sediment production from the gullies was 2480±80 t/ha/y which was declined in the period between 1958 and 1992 (1550±50 t/ha/y) mainly because of afforestation program to stabilize small gullies.

Burkard and Kostaschuk (1997) examined the change in area and length of gullies in the eastern shoreline of Lake Huron, Canada using sequential aerial photographs between 1930 and 1992. They found that larger gullies with larger watersheds had higher area growth rates, however, smaller gullies were found mostly stabilized. They attributed the cause of increase in growth rates to the extreme flow events, extension of municipal drains and use of sub-surface drainage.

Nachtergaele and Poesen (1999) used high-altitude stereo aerial photographs to estimate soil loss from ephemeral gully erosion in central Belgium. They studied the newly formed gullies as a result of intense rainfall using aerial photographs as well as field measurement.

They found that neither the aerial photograph nor the field survey could trace all the gullies, indicating the fact that combination of the two yielded a better result. They compared the two techniques comprehensively in the assessment of gully erosion. Main advantage of aerial photographs is time saving and cost effectiveness. Also, aerial photos permit the ephemeral gully erosion survey to be extended in time. Main drawback is that small gullies are usually invisible. Contrary to aerial photo method, high accuracy is possible by field survey. However, field measurement is costly and area or number of gullies to be covered is limited.

Throughout Nepal Himalaya, gully erosion is a widespread form of land degradation which is highly sensitive to climate and land-use changes (LRMP, 1986). Yet, research on gully erosion is very limited. Higaki et al. (1998) carried out a study on gully erosion in a fluvial terrace slope called "Pipaltar" in the Mid-Hills of Nepal. Long-term monitoring of head retreat was undertaken in three gullies formed on the slopes composed of laterite soil. They found that average rate of retreat decreased as a result of applied countermeasures in the form of gabion check dams and plantation. Crack formation was evident in the dry periods and retreat by block collapse occurred mostly during the rainy period.

Characteristics of study gullies

Specific erosion processes in the gullies were inventoried by means of geomorphological mapping of the gully head and sketches of characteristic gully head profiles. Gully sidewall morphology was traced from longitudinal and cross-section profiles prepared after levelling survey. Changes in vegetation cover on the headwalls were noticed by means of photo monitoring, which has been carried out for the last 5 years. It was also useful for understanding many relevant geomorphologic processes and erosional activities such as rill formation, scour channel formation, failure through block collapse, temporary deposits of sediments and so on.

A wide range of gully exists in the study catchment in terms of size and geometry, and their erosional activity varies widely. In view of the importance of the large active gullies in generating significant sediment in a catchment, the study is concentrated on the development mechanism of such gullies. They are characterized by actively eroding tall headwalls with minimal or no vegetation cover and significant amount of debris deposited in the fan area. Three such gullies namely Khjuri-1 (KG1), Khajuri-2 (KG2) and Musahar-1 (MG1) were selected for field monitoring (Fig. 8.8, refer Fig. 3.2 in Chapter 3 for the location).

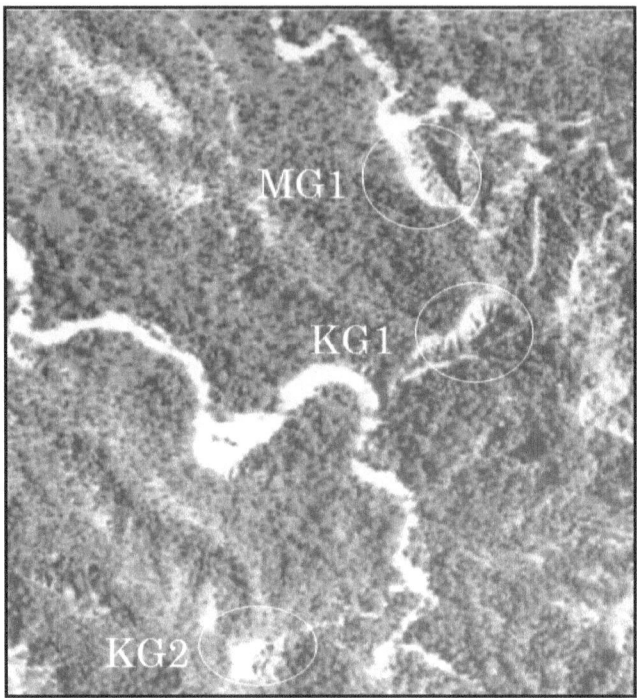

Figure 8.8 Study gullies traced on the 1992 aerial photograph.

The gullies are fully developed ones whose headscarps are nearly at the hill-crests. Gully head area typically shows leaf-shaped feature, which is much bigger in size compared to the planform area of the main channel. Many micro-channels cutting the head area can be visible from the aerial photographs, which can be a basis for identifying such gullies in the study area. Their rugged and virtually bare form reveals the fact that compared to the main channels; significant amount of sediment must be supplied from the erosion of gully heads.

Fig. 8.9 illustrates an example of gully-stream coupling in the hills. The gully head are fully developed which cut into the spurs running perpendicularly to the main ridgelines. In the head area, the slope of the headscarp is much steeper than the original hillslope forming a near-vertical face cliff which is coupled with the main channel through many drainage channels. The head is then linked to the main stream by a main channel which receives runoff and sediment from the gully head as well as from micro-channels formed on the non-gullied hillslopes. Additional instabilities such as channel bank scouring and isolated landslides can aid the sediment to the gully channel. The gully and mainstream channels are linked by a debris fan, which is a temporary storage of sediment.

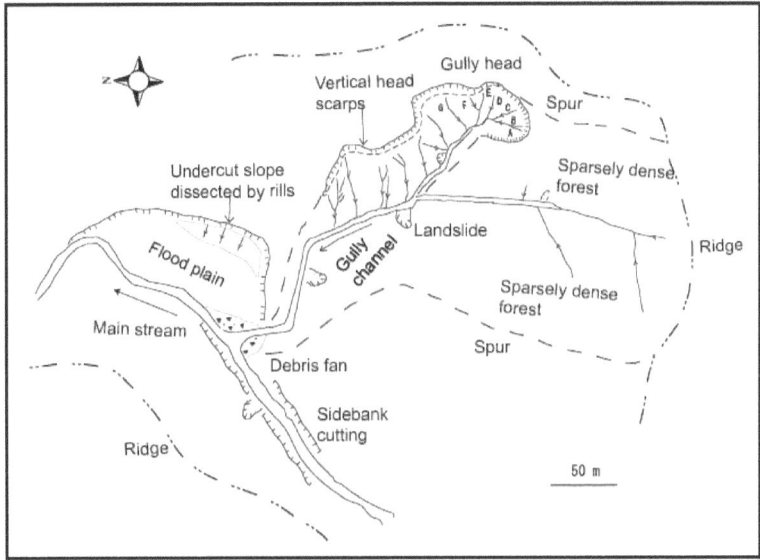

Figure 8.9 Gully plan features (KG1) along with gully-stream network (Modified after DPTC, 1998).

Salient features of the study gullies are presented in Table 8.4. Gully head is usually less than one hectare in area. Gully heads are typically high (maximum height up to 50 m) and are formed in different geological beds, mainly consisting alternate layers of sandstone and mudstone with thick intermediate layers of gravel and boulder (DPTC, 1998). They consist of nearly vertical faces which are dissected by numerous rills. When the rills coalesce downstream, a series of drainage channels are formed on the foot of headwall. So, the head distinctly consists of four components: head-rim, free-face, drainage channel and pediment slopes on the foot of channels (Fig. 8.10 and Fig. 8.11). The drainage channels, which are v-shaped in cross section, are composed mainly of unconsolidated gravel and boulder layers. Side slopes of the channels are almost bare which indicates that they are in active stage of erosion (Fig. 8.12).

Main channels are characterised by very steep slopes with the slope gradient varying from 12 to 18% (Fig. 8.13). However, they are relatively short and narrow, and sideslopes are covered with good vegetation, even though sporadic bank undercutting and small slope failures occur in some sections. It is important to note here that because of apparently smaller contribution of sediment compared to that from the head area, the main channels have been discarded in the estimation of the sediment yield from such type of large gullies. It must be cautioned that this assumption may not be valid for other types of gullies, which have relatively more unstable channels contrast to the head areas.

Table 8.4 Morphometric characteristics of the study gullies

Parameters	KG1	KG2	MG1
Gully Length, m	280	263	125
Average width, m	13.8	8.5	7.0
Average gradient, %	17.8	12.6	14.0
Area of Gully head (1992), ha	0.60	0.36	0.88
Average height of headwall, m	36	32	27
Area of alluvial fan, ha	0.25	0.16	0.19

Another basis of identifying the active gullies is the existence of debris fan at their outlets (Fig.8.14). Field observations indicated that the level of fan fluctuates frequently as a result of intermittent supply of debris from the gullies upstream. The debris fan being generally higher than the bed level of the main stream repulses the stream flood towards the opposite banks so that the banks are characterized by slope instabilities followed by toe undermining. The sediments deposit temporarily on the fan area which consists of large boulders mostly varies from 3 to 12 cm (Fig. 8.15).

Gully head enlargement by aerial photographs

Gully heads were traced from the rectified aerial photographs for examining the rate of enlargement. For this, the aerial photographs of 1964, 1978 and 1992 were used. The satellite image of 2003 which has a less resolution than these photographs was not used for this analysis because head rims of the gullies cannot adequately be identified on the image. Two variables were investigated: area and retreat rate of gully head. Growth rates were calculated by measuring the difference in gully area or retreat over the time period between successive aerial photographs, then dividing this by the number of years between the photographs.

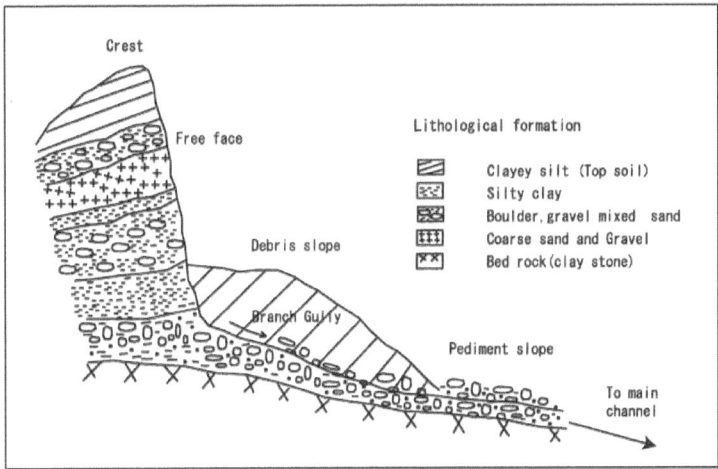

Figure 8.10 A schematic illustrating different segments and lithological formation of a gully head. The height of the free face cliff is about 36m at Gully KG1.

Gully-heads traced from the rectified aerial photos were analyzed for the head area enlargement rate. Area and distance were measured by using digital planimeter. Fig. 8.16 depicts the tracing sketches of head-rims of the gullies from the sequential photographs and Table 8.5 summarizes changes in gully head area and retreat rates for the studied periods.

Figure 8.11 Gully head of KG1.

June, 2002   September, 2002   August, 2003

Figure 8.12 Photo comparison of an active drainage channel (KG1-C). Significant scouring is evident on the sidewalls. It also depicts the temporary storage of debris materials on the channel bed.

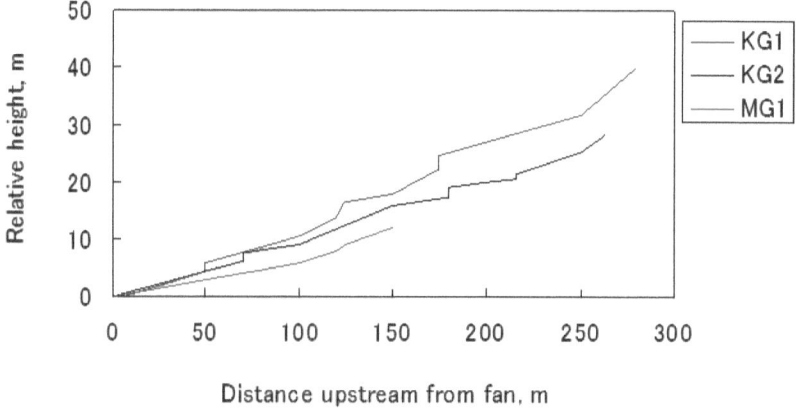

Figure 8.13 longitudinal sections of the main channels.

Figure 8.14 Formation of a debris fan of gully KG1.

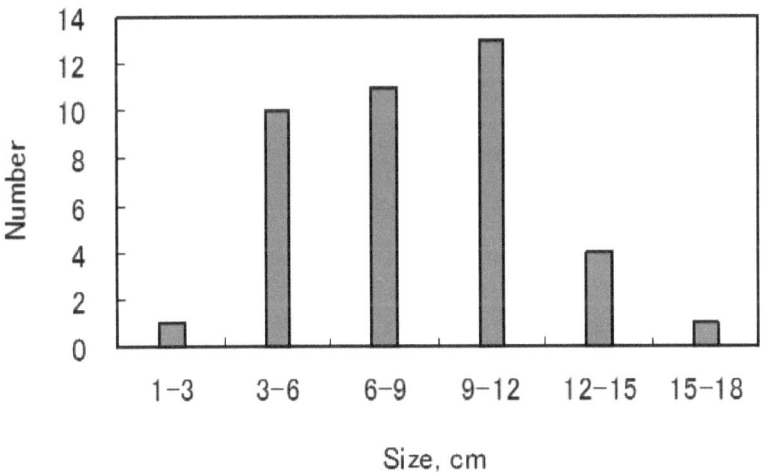

Figure 8.15 Maximum size of boulders around the debris fan of gully KG1.

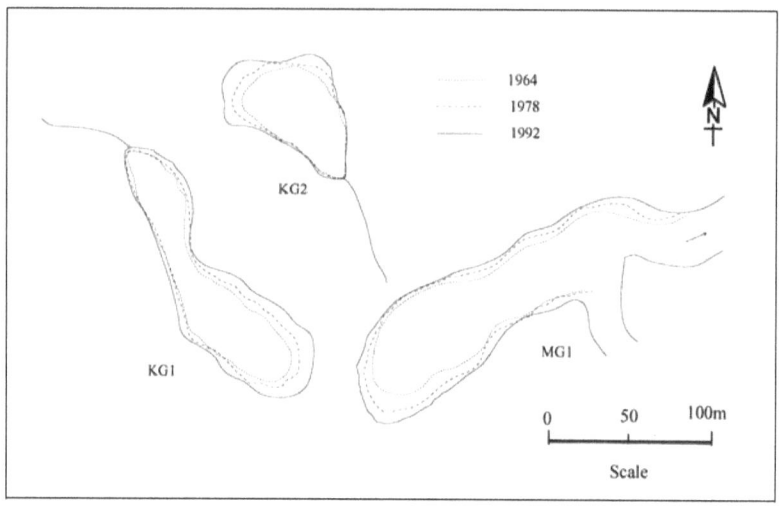

Figure 8.16 Head area enlargement of study gullies by aerial photo interpretation.

Table 8.5 Increase in gully head area and head retreat from 1964 to 1992.

	Year/period		KG1	KG2	MG1
Head area	1964	m²	3842	2436	6575
	1978	m²	4672	2896	7942
	1992	m²	6059	3555	8801
	1964-1978	%	21.6	18.9	20.8
	1978-1992	%	29.7	22.7	10.8
	1964-1992	%	57.7	45.9	33.8
Head retreat	1964-1978	m	2.9	5.9	12.7
	1978-1992	m	10.5	9.7	7.7
	1964-1992	m	13.3	15.6	20.4
	Av rate (28 yrs)	m/yr	0.48	0.55	0.73

Fig. 8.16 indicates that head area enlargement rate varied in space and time remarkably. Maximum head retreat occurred in headward direction, however sideward expansion was also noticed. Table 8.5 reveals that head area of the gullies increased by about 19 to 22% from 1964 to 1978, and the areas further increased by 11 to 30% from 1978 to 1992. Thus, in the whole period of 28 years, KG1 increased by 58%,

KG2 by 46% and MG1 by 34%. Total head retreats of 13.3 m, 15.6 m and 20.4 m occurred in the whole period. This resulted in the average rates of head retreat, which were 0.48, 0.55 and 0.73 m per year for the three studied gullies respectively.

Measurement of gully head erosion

Gully head retreat was measured as the change in distance between the edge of the gully head and benchmark pins fixed at the rim of the gully head. The pins are fixed at an interval of 15 to 25 m along the periphery of the head rim. Erosion from channels was measured by inserting iron pins (1 cm diameter, 30-50 cm long) perpendicular to the slope surface and repeatedly measuring the exposed segment, as described in Chapter 3. The pins are fixed at an interval of 3 to 6m on the sidewalls.

Based on the form and process, erosion sources were divided into two types: retreat of vertical face (headwall erosion) and drainage channel erosion. In order to look at the retreat pattern during rainy season (from June to September), it was planned to carryout field measurements before and after the rainy season as well as during the dry season. Thus, frequency of monitoring was at least two times a year. Additional field observations were undertaken during the mid-rainfall season in order to closely observe erosion processes such as overland flow erosion and mass failures.

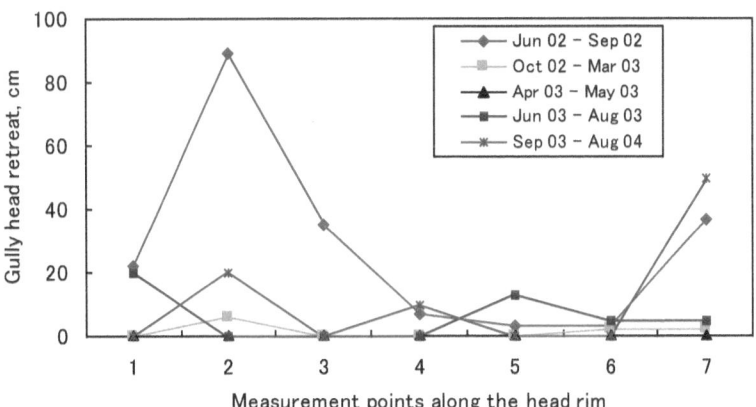

Figure 8.17 Gully head retreat amount in different time periods at gully KG1. The numbers along x-axis denote the positions of measurement along the head rim. Generally the distance between the measurement points varies from 15 to 25 m (Refer Fig. 8.12 for more explanation).

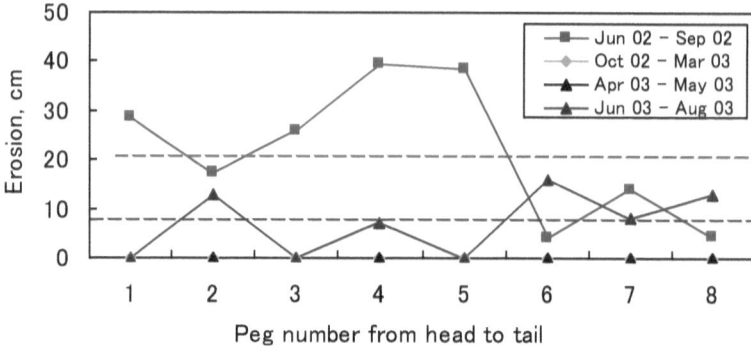

Figure 8.18 Channel sidewall erosion amount at an active drainage channel (KG1-C). The peg number indicates the position of the measurement, the measurement point 1 stands for the head and 8 for tail, with an interval of about 3 m. Refer Fig. 8.12 for location of measurement points.

Emphasis was given to the monitoring in the rainy season because the area gets more than 70 percent of annual rainfall in this period. It should be noted here that head retreat of only two gullies- KG1 and MG1 could be measured because gully head of KG2 was inaccessible. For the measurement of channel erosion, one representative channel from each gully was selected. As example, head retreat and channel sidewall erosion measurement at gully KG1 are shown in Figs. 8.17 and 8.18 respectively. Erosion measurements undertaken over different periods are presented in Table 8.6.

The table indicates that gully walls did not retreat uniformly along the perimeter. There was also a considerable temporal variation in erosion amounts. From June through September 2002, maximum headwall erosion occurred was 89cm and 30cm in KG1 and MG1 however, they were 26 and 9 cm in the same period of 2003. Similarly, there was considerable temporal and spatial variation in channel erosion. It also indicates that there was virtually no erosion during winter or dry seasons.

In spite of the wide variation in the erosion rates, it is found consistent that headwall as well as channel wall erosion across the three studied gullies were higher in 2002 compared to that in 2003 (Fig. 8.19). Even though there can be a host of factors to affect the erosion patterns, difference in erosion amounts in the two years can partially be explained by the difference in rainfall pattern. The monthly rainfall in July 2002 was 680 mm whereas it was 490 mm in the same month in 2003. However, effect of rainfall size and intensity on the erosion pattern was not very clear.

Table 8.6 Measurement of headwall retreat and channel side slope erosion (unit in cm).

Gully			02-Jun-02 30-Sep-02	1-Oct-02 15-Apr-03	16-Apr-03 06-Jun-03	07-Jun-03 25-Sep-03	26-Sep-03 16-Aug-04
KG1	Headwall n =7	Min	3.0	0.0	0.0	0.0	0.0
		Max	89.0	0.0	6.0	20.0	50.0
		Mean	**28.0**	**0.0**	**1.4**	**6.0**	**10.0**
	Channel n = 8	Min	4.0	0.0	0.0	0.0	-
		Max	40.0	0.0	0.0	16.0	-
		Mean	**22.0**	**0.0**	**0.0**	**8.2**	-
KG2	Headwall		-	-	-	-	-
	Channel n = 13	Min	3.5	0.0	0.0	0.0	-
		Max	28.0	0.0	0.0	11.0	-
		Mean	**8.8**	**0.0**	**0.0**	**2.9**	-
MG1	Headwall n = 6	Min	2.0	0.0	0.0	0.0	0.0
		Max	30.0	3.0	0.0	9.0	15.0
		Mean	**16.0**	**1.0**	**0.0**	**4.0**	**3.0**
	Channel n = 10	Min	20.0	0.0	0.0	6.0	-
		Max	35.0	0.0	5.0	20.0	-
		Mean	**27.0**	**0.0**	**1.0**	**10.3**	-

Note:

n: number of erosion pins, Min and Max: Minimum and Maximum erosion, Mean: Mean of all erosion pins. " - " stands for "not measured".

During the monitoring in August 2004, erosion of channel sidewalls could not be measured due to inaccessible condition. However, it was possible to measure the head retreats of KG1 and MG1. In the period between September 2003 and August 2004, the average erosion amount for these gullies is found to be 10 and 3 cm respectively. As explained in Chapter 5, much bigger rainfall events occurred in the month of June in 2004. However, the erosion amount (head retreat) during this period is less than in 2002. It indicates that major rainfalls not always induce more erosion, which may need some additional events to extend the erosion up to the measurement points at the head rim.

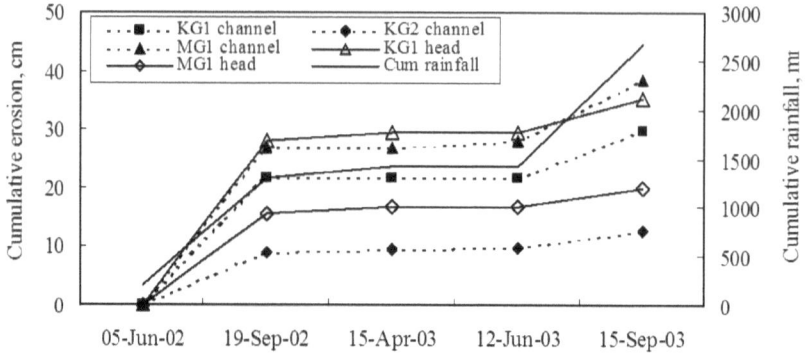

Figure 8.19 Erosion verses rainfall in 2002 and 2003.

Estimation of sediment production

Sediment volume derived from head retreat has been calculated in two ways: from aerial photography and field measurement. From aerial photography, difference in gully head area is calculated from 1964 and 1992 aerial photographs and total sediment volume is calculated by multiplying this difference by average gully head height. Height of the gully head is taken as the distance between the gully head rim and a level below which sloping channels are formed, and it is assumed to be constant as it may increase when gully retreats upstream. Average height was calculated by averaging the heights measured from different points on head-rim. Channel erosion is excluded from this estimation because change in channel forms and dimension was difficult to detect from the photographs. Calculation of sediment volume is shown in Table 8.7.

Table 8.7 Estimation of sediment production from gully head using aerial photographs.

Gully	Area 1964 m²	Area 1992 m²	Diff m²	Av height m	Total vol m³	Vol/year m³/yr	*Wt/yr t/yr	Av. Rate t/ha/yr
KG1	3842	6059	2217	32	70944	**2534**	3547	5854
KG2	2436	3555	1119	24	26856	**959**	1343	3778
MG1	6575	8801	2226	35	77910	**2783**	3896	4427

* Bulk density considered as 1.4 t/m³

From field measurement, eroded sediment volume was calculated for the two monitoring years by considering also side slope erosion from the channels. Headwall erosion was estimated by measuring headwall surface area and multiplying it by average measured erosion depth. Eroded volume from channel side slopes were estimated by measuring channel length and slant height of sidewalls, which were then multiplied by average erosion depth. Repeated survey indicated that channel beds fairly maintain the longitudinal slope because debris derived from the headwall and channel sidewall temporarily protects the bed. Hence, sediment production from the bed is not considered. Because channel forms and material properties do not differ much in the gullies, average channel dimensions could give reasonable estimates. Sediment volume estimation is presented in Table 8.8. The erosion amount in 2004 is not included in the estimation because of limited measurements and lack of yearly rainfall data.

The table indicates significant differences in total eroded volume in the two years. A total erosion of 2793 m^3 in KG1 and 2572 m^3 in MG1 took place in 2001, however it decreased to 738 m^3 and 731 m^3 respectively in 2002. It also illustrates that contribution of channel erosion to total erosion is less than that of gully head erosion.

Table 8.8 Estimation of sediment production by field measurement.

Parameters	2002			2003		
	KG1	KG2	MG1	KG1	KG2	MG1
Headwall erosion						
Length of head, m	273	208	359	273	208	359
Height of head, m	32	24	35	32	24	35
Av retreat, m	0.28	~	0.16	0.07	~	0.04
Volume, m³	2446	~	2010	612	~	503
Channel side slope erosion						
No of channel, N	9	12	11	9	12	11
Av length, L m	25.0	18.0	21.0	25.0	18.0	21.0
Av slant height, H m	3.5	3.0	4.5	3.5	3.0	4.5
Av erosion, D m	0.22	0.09	0.27	0.08	0.03	0.11
*Volume, m³	347	117	561	126	39	229
Total eroded volume, m³	2793	~	2572	738	~	731
**Total eroded weight, tones	3910	~	3600	1033	~	1024
Gully head area, ha	0.60	0.36	0.88	0.60	0.36	0.88
Erosion rate from gully area, t/ha	6517	~	4091	1722	~	1164
Average annual rate, t/ha/yr	4120 for KG1			2628 for MG1		

* Volume = 2*N*L*H*D for two side slopes
** Bulk density of sediment as 1.4 t/m³

Discussion

Erosion productions estimated by the two methods differed significantly. In aerial photography, only headwall retreat was considered as a source of sediment. Meanwhile channel sideslope erosion was also considered in field measurement method even though its contribution to the total sediment production seemed less. If only sediment volumes from headwall estimated by the two methods are compared, they would differ by 8% for KG1 and 80% for MG1 in 2002. As the measured erosion was consistently less, the difference would be even more for 2003. There could be a number of possible factors to explain this discrepancy. Average retreat rates derived from long-term intervals mask seasonal effects or other period of stagnation or activity, but do so persistent trends (Oostwoud Wijdenes and Bryan, 2001). In the short term, retreat rates show much variation since factors other than runoff such as development of tension cracks, deposition of sediment below the headcut etc become increasingly significant. Many studies such as carried out by Vandekerckhove et al. (2001) indicated less variability in long-term retreat measurements than in short-term ones. In our case, variation in erosion amounts in the two years was significant, possibly due to the difference in rainfall pattern as explained before. However, it would be less in case of long-term period because rainfall characteristics may average out over time.

Retreat of headwall is often referred to the shifting of head-rim upstream. However, erosion may occur at the foot of the headwall while this undercutting may not result in collapse of the complete wall, so no retreat is measured. This type of partial block collapse was evident in many locations of gully heads. It may be possible that the propagation of the mass failure on the headwall up to the head-rim may take longer time or require several storm events especially in high heads.

In most of the studies related to sediment production by headcut advance, rainfall is used as a predicting parameter, however, explicit relationship between rainfall amount and head retreat rates is often not clear. Oostwoud Wijdenes and Bryan (2001), for example, found no apparent relationship between rainfall amount and gully head retreat rates from 32 storm events. They attributed this poor correlation to the differential erosion of the head. This could particularly be true for the high gully heads. In calculating the sediment production it was assumed that retreat occurred over the whole contour of the gully head. But measured retreat rates differed widely along the head-rim. However, the profile of the head-rim may remain fairly constant and the assumption could be valid for a long-term period. It suggests that only two-year data would not be sufficient to explain the long-term sediment production rates. The monitoring data thus simply represent a short-term erosion process.

Figure 8.20 Crack formation on the head-rim of gully MG1.

Geological composition of the headwall makes the erosion process too complex, which consists of many strata of different formations. The headwall, for example, consists of alternate beddings of unconsolidated silt, sand and boulders derived from siltstone, sandstone and conglomerate. At saturation, selective washout of silt and sand induces free-fall of boulders, which ultimately leads to mass failure. Cracks at the head rim are the indicators of the potential failure (Fig. 8.20). Formation of crack in the dry season followed by block collapse in the rainy season was also noticed by Higaki et al. (1998) in the gullies formed on the laterite of fluvial terrace in Midlands. The collapsed materials fill up the plunge pool and bed of drainage channels temporarily, which can to some extent contribute in checking the undercutting of drainage channels. Since the channels are very steep, this effect lasts for a short time until a few rainstorms occur. This process may result episodic mass movements on the headwall and drainage channels.

Summary

Following points can be summarised:
- Long-term change analysis by sequential aerial photo comparison indicates that the gullies expanded remarkably over the period between 1964 and 1992 by 34 to 58% of the area in 1964. Maximum retreat rates during that period for KG1, KG2 and MG1 were 0.48, 0.55 and 0.73 m/year respectively. Estimated eroded volumes were 2534, 959 and 2783 m^3/year from respected gullies.

- From field measurement, eroded volumes estimated for KG1 and MG1 were 2793 m^3 and 2572 m^3 in 2002, and 738 m^3 and 731 m^3 in 2003.
- Difference in estimated volume from the aerial photography and field measurement could be due to the fact that average retreat rates derived from long-term intervals mask seasonal effects. Also, there could be differential erosion along the head-rim as a result of rainfall in a short-term period, however the effects of rainfall characteristics on headwall erosion may average out in the long-term period.
- Annual average erosion rates from the gullies vary from about 2600 to 4100 t/ha. It indicates that the gullies are the important point source of sediment in the catchment.
- The study has outlined the complexities in headwall and drainage channel erosion pattern especially in multi-bed geological formation. Also, dominance of mass failure in the erosion process has been recognized. Sediment production from these gullies seemed episodic in nature depending on stage of development and rainfall patterns.

Landslide slope erosion

Soil loss by mass movement is the most significant type of erosion in the steep hilly landscapes (Leopold et al. 1964; Morgan, 1986). Many studies have suggested that sediment contribution from mass movement in the steep hill catchments is significantly higher than from surface erosion (eg. Merz, 2004; Morgan, 1986; Selby, 1993). Although mass movement like landslide has widely been studied in Nepal, it is generally neglected in the context of sediment production.

In the Siwalik Hills, mass movements occur in various forms, however in the present context, only landslides are taken into account because sediment production from them could be substantial. Mass movement activities have been noted in various studies (eg. Upreti, 2001; Sharma, 1981, WECS, 1987) however no quantitative information is available on the rate of sediment production from landslides.

Landslides in soil can broadly be divided into two types: (i) Translational landslides and (ii) Rotational landslides (Varnes, 1958). Translational slides are by far the most common form of landslide occurring in soil (Selby, 1993). They are always shallow features, and have essentially straight slide planes which usually develop along a boundary between soil materials of different density or permeability. Shallow, rapid landslides typically occur on steep slopes and are often triggered by individual rainstorm events (Dhakal and Sidle, 2004). Rotational landslides, on the other hand, have curved failure plains and involve rotational movement of the soil mass.

Both types of slides exist in Siwalik Hills. Translational types mostly occur in the hillslope where relatively coarse grained materials such as gravel and boulder mixed soil and conglomerates are rich (Fig. 8.21). In the hill slope reach where the streams experience frequent meandering, slope failures exist as a continuum. The slopes are highly incised by the rills and to some extent they can form a badland landscape by excessive erosion.

Rotational types of landslide occur in the steep slopes composed of soft rocks such as shales and mudstones. They mostly occur in the toe of a slope which is undercut by stream flow (Fig. 8.22). The head of the displaced mass would be much closer to the crown of the slide and the length of the exposed slide plane would be small compared to the shallow slope failures (Selby, 1993).

It is important to note that for estimating rate of sediment production, no discrimination has been made between these two categories of mass failures. It is because, erosion process operating on the sliding slope surface has been assumed similar, since they develop on similar physical environment where toe undercutting is the main triggering factor.

Figure 8.21 Translational landslide at the side wall of a tributary channel.

Process of landslide slope erosion

Landslides are often event-driven phenomenon. In the local scale, sediment production from them is the total volume of moving mass which is easier to compute with the help of direct field measurement. When such events are considered in the catchment scale, erosion process operating on the slide surface is required to be ascertained which is not easy however. It is mainly due to the fact that erosion rate varies widely in relation to the type and rate of landslide movement. In addition, erosion resistance largely depends on the degree of deformation of soil material and water content of sliding mass (Selby, 1993).

Since occurrence of landslide is a threshold process, initial sediment production (volume of sliding mass) is an event rather a gradual process. Field observation revealed that most of the active landslides loss the sliding mass within some years as a result of successive erosion and active down cutting by runoff flow, or partly of it can be retained on its foot slope. However, the slide surface continues to be eroding for many years until it is stabilised by a recovery of vegetation. For this reason, landslides are not only mass wasting events but also phenomenon of gradual erosion.

Figure 8.22 A Rotational landslide at the sidebank of Khajuri stream triggered by toe cutting.

From this viewpoint, there are two broad process phases in landslide as regards sediment production. One is the production of sliding mass under gravity just after the occurrence, a process of mass wasting, and other is the persistent erosion of head scarp and exposed slide surface. While estimating the sediment production rate from the landslides, only erosion from the slide surface has been considered in the present study.

It is obvious that sediment production from landslides largely depends on the state of activity. Fresh landslides suffer extensive erosion on the head scarp and sliding surface while erosion capacity of dormant landslides is low. Vegetation density is a key indicator of erosion activity on sloping surfaces (Crouch and Blong, 1989). In this study, two types of landslides are distinguished based on erosion activity:

(i) Active landslides: They are recently occurred landslides with active erosion operating on the sliding surface and,
(ii) Stabilized landslides: They are old landslides often rich in vegetation with minimal erosional activity.

Only active landslides have been considered for estimating erosion rate.

Mapping and field measurement of landslides

Landslide mapping was done using aerial photograph and field verification (Fig. 8.23). Three active landslides namely L1 and L2 and L3 (Location shown in Fig. 3.2; Chapter 3) were monitored for slope erosion using erosion pins. Table 8.9 shows the characteristics of study landslides and average erosion rates.

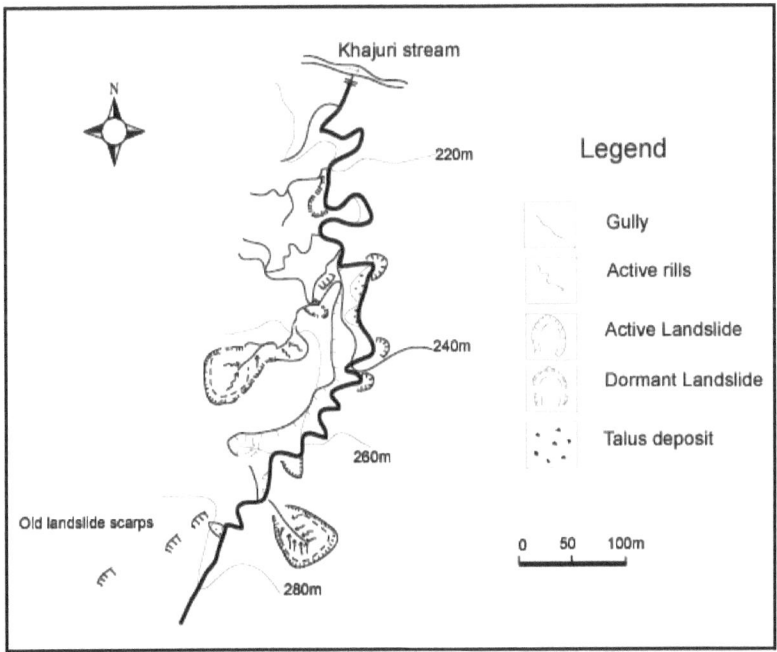

Figure 8.23 Field mapping of some landslides in the Khajuri catchment (After DPTC, 1988).

Table 8.9 Characteristics of landslides and average erosion rates (May, 2002-May, 2003).

Landslide	Slope area, m²	Slope degree	No. of pins	Erosion cm	Av. erosion cm
L1	1200	50	3	9	
				6	6 ± 3
				4	
L2	980	45	2	7	
				5	6 ± 1
L3	645	40	2	6	
				4	5 ± 1

In Table 8.9, erosion amounts are the average erosion at different sections along the slide surface. It indicates that erosion rates vary from 4 to 9 cm a year. The erosion rates are much higher than the surface erosion on a bare slope surface apparently due to the fact that landslide slopes are covered by loose materials often transported from upslope. It was observed that landslide erosion especially on the foot slopes takes place predominantly by rill erosion while head scarp erosion takes place by mass wasting. In fact, the erosion rates given in Table 8.9 represent erosion from rills formed on the sliding surface of the landslides.

Summary

Following points can be summarised:
- The steep hillslopes exhibit a large density of landslides especially on the foot slopes mainly induced by stream channel undercutting.
- Vegetation density is a key indicator of erosion activity on the sloping surfaces. Both active and stabilised landslides occur in the area.
- Rill erosion is a predominant process occurring on the slide surface slopes.
- The average erosion rates vary from 4 to 9 cm/year.
- It is important to note that the technique used here to estimate the soil loss from a landslide is only approximate. The landslides and mass wasting phenomenon are in fact very complex processes which generally do not occur on a uniform fashion.

9

Streambank erosion

Background

River bank erosion can present a wide variety of environmental problems through the loss of forest and agricultural land, danger to riparian and floodplain structures, increased downstream sedimentation and occasional riverine boundary disputes. River bank erosion is a process of both geomorphological and anthropogenic significance and these aspects are apparent over different time-scales (Couper and Maddock, 2001). Bank erosion processes are the key components of meander formation, lateral channel migration and movement of sediment throughout the drainage basin (Lawler et al., 1997).

Degradation of catchments and streams has led to increased sediment yield with consequent impact on water quality and aquatic ecosystem (e.g. Wasson et al., 1998, Walling and Webb, 1983, Sutherland and Bryan, 1991, Bunn et al., 1998). Segregation of sediment contribution from a wide range of sources is extremely difficult because of the complexity in the processes and controlling factors. However, many studies advocate on the role of bank erosion in the basin sediment yield. In humid to semi-arid regions significant amount of sediment load can be produced from bank erosion which have experienced historical land use intensification and degradation of valley-floor and riparian vegetation (Prosser et al., 2000). Such degradation has resulted in massive expansion of channel networks in the form of incised streams and gullies often eroded into ancient colluvial and alluvial deposits (Graf, 1979; Schumm et al., 1884; Prosser et al., 1994). Bank erosion may contribute significant amount of material to overal catchment sediment yields (Imeson and Jungerius, 1977, Grimshaw and Lewin, 1980, Trimble, 1993, Church and Slaymaker, 1989, Bull, 1997, Simon, 1996). The contribution may vary from less than five percent (Walling and Woodward, 1992, Walling and Kane, 1984) to over 80 percent (Imeson and Jungerius, 1977, Imeson et al., 1984, Simona and Thorne, 1996). In spite of ongoing intensive studies on bank erosion process, a large gap still exists in understanding the relationship between bank erosion and sediment dynamics (Bull, 1997).

Measurement techniques

A large number of techniques are available to measure bank erosion. Laeler et al., 1997 identified eight major methods depending upon the timescales of application, which include:

(a) long-timescale techniques- sedimentological evidence, botanical evidence and historical sources
(b) medium-timescale techniques- plannimetric survey and repeated cross-profiling.
(c) Short-timescale techniques- erosion pins, terrestrial photogrammetry and the Photo-Electronic Erosion Pin (PEEP) system.

The measurement technique should be selected depending on the study aim, logistical constraints and site conditions. In the present study, method of erosion pin is applied which is briefly outlined here.

Erosion pin technique

Erosion pin deployment is the most popular technique often used for short time-scale aiming to identify causes, processes and mechanisms of bank erosion. The method is simple and cheap to use which can pick up temporal and spatial patterns of bank erosion in a reasonably wide range of fluvial environments. Many studies on erosion measurement have employed this technique (e.g. Gardiner, 1983; Ashbridge, 1995; Lawler et al., 1997, 1999; Stott, 1997; Couper at al., 2002).

The basic principle of the erosion-pin technique is that a pin is inserted into the bank, leaving a known length exposed to provide a `benchmark` against which bank erosion can be measured. Any increase in exposure of the pin is taken to represent erosion of the bank. The pins are often made of metal, between 4 to 10mm in diameter and 25 to 50 cm long.

In spite of its simplicity and wide range of applicability, the method has limitations as well. Couper et al. (2002) has pointed out some problems. First, many disturbances such as excessive scouring can cause the loss of the pins. Second, it is possible that the presence of erosion pins themselves or the act of inserting them into the bank may disturb the processes in operation. Third, pins can be disturbed or sometimes lost by human interference.

Bank erosion: a complex and diverse fluvial process

Complexities in the bank erosion process have widely been outlined. Bank erosion takes place with complex interaction of hydraulic and geotechnical forces (Simon et al., 2000) and it displays a wide variety of spatial and temporal patterns. Even within the same hydroclimatic environment, bank erosion rates vary widely with river system scale, channel geometry, bank material characteristics and in response to other channel changes taking place (Lawler et al., 1997). There is a large variety of bank erosion mechanisms and causes, and processes act in concert to produce bank retreat. The combination themselves are likely to vary at seasonal and subseasonal timescales. Most of the recent studies emphasize the interplay of a variety of weakening, fluvial erosion and mass failure processes in producing serious bank erosion, instability and /or retreat (refer Box 9.1 for definition of terms).

Process of bank erosion

Three primary mechanisms are involved in riverbank erosion: subaerial weakening and weathering, fluvial entrainment and mass failure. Subaerial processes, which include wetting and drying of the soil, associated desiccation and freeze-thaw activity, are commonly thought of as `preparatory` rather than erosive processes (e.g., Throne, 1990; Lawler et al., 1997; Green et al, 1999). However, Couper and Maddock (2001), Couper et al., (2002) demonstrated that the erosion processes operate at different levels of magnitude and frequency, and role of sub-aerial erosion can be significant implying that they can also be erosive agent themselves.

The second type of mechanism- fluvial entrainment is related to removal of bank material as individual grains or aggregates by hydraulic action of flowing water. Erosion processes greatly differ in the bank composition, which can be divided into two types: cohesive and non-cohesive banks. Cohesive, fine-grained bank material is usually eroded by the entrainment of aggregates or crumbs of soil rather than individual particles which are bound tightly together by electromechanical cohesive forces. Because of this complex property, erosion of cohesive stream banks is more complex to analyze. No complete theory of this type of erosion exists (Lawler et al., 1997). Non-cohesive materials, on the other hand, are detached and entrained grain by grain because of lack of electrochemical bonding. Stream banks composed of cohesionless soils normally are highly stratified heterogeneous deposits. Bank erosion is controlled by gravitational forces and particle characteristics.

Box 9.1 Definition of terms relevant to bank erosion (Lawler et al., 1997).

- **Bank erosion**: Detachment, entrainment and removal of bank material as individual grains or aggregates by fluvial and subaerial processes.

- **Bank failure**: Collapse of all or part of the bank en masse, in response to geotechnical instability processes.

- **Bank retreat**: Net linear recession of bank as a result of erosion and/or failure

- **Bank advance**: The opposite of bank retreat, i.e. net linear stream wise change in bank surface position, as a result of deposition of sediment or in situ swelling of bank materials.

- **Bank erodibility**: The ease with which bank material particles and aggregates can be detached, entrained and removed normally by flow processes.

The third type of erosion mechanism is mass failure which can be analyzed in geotechnical slope stability terms. When a section of a bankline fails and collapses, blocks of bank material slide or fall towards the toe of the bank. They may remain there until broken down in situ or entrained by the flow. Cohesion of the failed material can provide additional strength to resist direct fluvial entrainment. Failed cohesive materials can act as a form of natural bank-toe protection by consuming and diverting flow energy (Wood et al., 2001; Simon et al., 2000). Box 9.2 lists out key controlling factors of bank erosion.

Vegetation and bank erosion process

Bank erosion processes and failure mechanisms act in different ways to produce bank retreat. Vegetation can significantly affect either or both facets of retreat in any particular stream bank. Compared to an unvegetated slopes covered by a good stand of close-growing vegetation experience an increase in erosion resistance of between one and two orders of magnitude (Kirkby and Morgan, 1980). Many studies have generally shown the positive overall effects of vegetation on bank erosion (e.g. Gray and Leiser, 1982; Stott, 1997; Simon and Collison, 2002; Micheli and Kirchner, 2002). However, Thorne (1990) has noted that bank vegetation can be both beneficial and detrimental. The potential impact of vegetation on the bank stability is strongly related to their type, age, health, density, spacing and many other plant characteristics like rooting system. While grasses and shrubs are effective at low velocities, their impact decreases as velocities increase. Conversely, the stems of woody species continue retarding the flow up to very high velocities, but may generate serious bank scour through the local acceleration of flow around their trunks. So, spacing of the trees becomes an important factor. Single trees or small group of trees are impediments to the flow that generate large-scale turbulence and severe bank attack in their wakes. An isolated, downed tree may generate local scour and can lead to serious channel instability. In contrast, a dense accumulation of downed timber on a bank can be quite effective in protecting the bank from flow scour. The positive and negative effects of bank vegetation are given in Box 9.3.

Box 9.2 Controlling factors of bank erosion (Lawler et al., 1997).

- **Sub-aerial processes**: Microclimate- especially temperature, bank composition-especially silt-clay proportion

- **Fluvial processes**: stream power, shear stress, local slope, bend morphology, bank composition, vegetation, bank moisture content

- **Mass failure**: Bank height, bank angle, bank composition, bank moisture content, pore water pressure/tension

---

Issues of bank erosion

In Nepal, like other types, problem regarding bank erosion is widespread. The problem is more severe in the plain land of Terai where rivers frequently change their routes as a result of riverbed aggradations (Sharma, 1977, Lauterburg, 1993). However, research studies related to bank erosion process are very few. Most of them generally focus on the flooding and inundation problems, skipping the mechanism of bank erosion and channel stability. Issues regarding loss of fertile agricultural land and area of settlement are very common in the Siwalik Hills too (Fig. 9.1). However, very limited studies are available on the process of streambank erosion.

> **Box 9.3 Effects of vegetation on bank erosion (Gray and Leiser, 1982).**
>
> **Positive effects**
> - Foliage and plant residues intercept and absorb rainfall energy and prevent soil compaction by raindrop impact
> - Root systems(soil-root matrix) physically reinforce soil particles
> - Near-bank velocities are diminished by increased roughness
> - Plant stems dampen turbulence to reduce instantaneous peak shear stresses
> - Roots and humus increase permeability and reduce excess pore-water pressures
> - Depletion of soil moisture reduces water-logging
>
> **Negative effects**
> - Possibility of turbulence and local scouring
> - Increase in surcharge load
> - Weakening of bank strength by the roots of dead trees.
> - Trees may shade and suppress shorter vegetation that help to bind bank materials.

Objectives

In view of the apparent problems and potential source of sediment in the catchment, the main objective is to present a synthesis of results obtained from the intensive field measurements and observations of bank erosion rates, patterns and processes in the ephemeral streams. Followings are the specific objectives in relation to stream bank erosion:

(1) To determine the annual average rate of bank erosion
(2) To examine the pattern/processes of bank erosion
(3) To identify key controlling factors (particularly role of bank vegetation on stability of stream channels)
(4) To estimate sediment production from bank erosion in catchment scale and,
(5) To develop a methodology framework for bank erosion hazard evaluation

Stream bank characteristics

As evident by overlay of time-series aerial photographs, the streams have suffered by notable increase in width in most of the channel reach. It reveals the fact that severe bank erosion is prevalent in the stream channels. Field observations also supported this fact as many active cutbanks can be visible in many reaches of the streams. Salient features of stream banks commonly found in the streams are briefly explained below.

Figure 9.1 Examples of active stream bank erosion resulting in a loss of agricultural land.

Bank geometry

Most of the banks are typically vertical in profile. However, the banks composed of fine grained soil with high proportion of clay develop inclined profile. Bank height widely varies from 50 cm to as high as 15m. Height of majority of the banks ranges between 50 cm to 5 m.

Bank material

Based on material composition, streambanks can generally be divided into four types: bedrock, cohesionless banks, cohesive banks and stratified or inter-bedded banks. Of them, bedrock and stratified banks are prevalent in the study area. Banks composed of bedrock are associated with the uppermost reach (hills) of the streams. Majority of the banks formed on fluvial terraces are stratified. The soils in stratified banks consist of layers of materials with various sizes, permeability and cohesion. The degree of stratification is very high as a bank of merely 2m height displays four distinct unconsolidated inter-layered materials (Fig. 9.2). Uppermost 10 to 20 cm soil is fine silt or clay (black in colour), which is rich in organic materials. Layers below it contain the inter-bedding of fine materials (silt, fine sand) and coarse materials (gravel and boulder). Generally the bed layer is composed of silt or clay.

Figure 9.2 Inter-bedding layers of fine and coarse soil in a stream bank (respectively from top are: top black clay, gravel and boulders, silt and clay, scale shown- 1m).

Bank vegetation

Three types of bank-top can be found in relation to the near-bank vegetation: forest, shrub and bare. In the hills and upper terraces, majority of the banks consist of forest cover, especially young Sal (Shorea Robusta) forest. Partially dense forest and shrub cover dominate the lower terraces. No vegetation cover is remained further downstream exposing the banks to either agricultural lands or bare lands. Examples on bank vegetation are presented in Fig. 9.3.

Figure 9.3 Bank types in relation to vegetation cover: a. Bare , b. Shrubs, c. forested cutbank, d. forested stable bank.

Establishment of monitoring sites

Measurements of processes have been made on the most rapidly eroding sections rather than systematic sampling of all banks. However, relatively stable sections were also monitored frequently by means of repeated photography.

Altogether 10 bank sites were set up for erosion measurement. The sites were decided taking all types of banks into consideration in a bid to represent the key features of stream banks within the catchment. Characteristics of the selected bank sites are summarized in Table 9.1. The location of the bank sites is shown in Fig 3.2 in Chapter 3.

Measurement of streambank erosion

Using the technique of erosion pins as described above, monitoring of streambank erosion was undertaken since May, 2002 through August, 2004 covering three rainy seasons. Measurements have been made on the most rapidly eroding sections rather than systematic sampling of all banks. However, relatively stable sections were also monitored frequently by means of repeated photography.

Table 9.1 Characteristics of the selected stream banks.

Stream bank	# Material composition	Length m	Height m	Vegetation	Stream Width m	Gradient %	Bend angle degrees	no of pegs
KB1	C-G-B-Si-T	62	1.4	bare	80	1.6	110	55
KB2	C-G-B-Si-T	40	1.65	shurb	42.5	1.6	140	7
KB3	C-Si-G-B-L-T	33	3.2	forest	24	1.6	140	5
KB4	C-Si-B-T	62	1.5	forest	21	1.6	120	12
KB5	C-G-B-T	39	2.25	forest/roots	47	2.4	130	5
KB6	C-G-B-T	16	1.5	forest/roots	42	2.4	120	6
KB7	C-G-B-T	20	1.75	shurb	52	0.9	130	3
KB8	C-G-B-T	40	3.25	shurb	31	3.5	150	5
MB1	C-GB-Sa-T	70	6.5	bare	22	1.7	123	5
MB2	C-Si-L-B-T	12	7.5	forest	24	1.5	120	10

Material composition (from bottom to top): C- Clay, G- Gravel, B-Boulder, Si- Silt, Sa- Sand, T-Top soil (silt/clay)

* K- Khajuri stream, M - Mushahar stream

Erosion pins were inserted at a height 10 to 20 cm from the stream bed level where maximum erosion can be expected. The pins were kept at 5 to 20 m interval depending on the bank characteristics. The bank-top line was monitored in some banks by means of offset measurement from reference objects such as matured trees, boulders. In addition, a photo monitoring study was undertaken to find if there was any remarkable changes on the bank face. Average rates of erosion from each bank during the monitoring period are given in Fig. 9.4. The figure reveals that erosion rates vary from few centimeters to as high as 2 m a year. At KB2, the rate was highest mainly as a result of the scouring of the bank toe following inundation during a peak rainfall period (June-July) of 2002. Erosion was found minimum at the bank KB3. Erosion at the other banks was approximately 12 cm a year in average.

Temporal and spatial variation

It was found that there was notable variation in erosion amounts both in space and time which reflected the complex behaviour of the process under the combined effects of many controlling factors. At MB1, for example, both positive and negative pin erosion was observed (Fig. 9.5). The reason is that many of the pins were covered by the debris fallen from the bank. Couper et al. (2002) describes such phenomena which may result from the deposition of sediment by high flows or mass failure of the bank. It indicates that the erosion amount varies widely in relation to the position of the pins in the vertical bank profile.

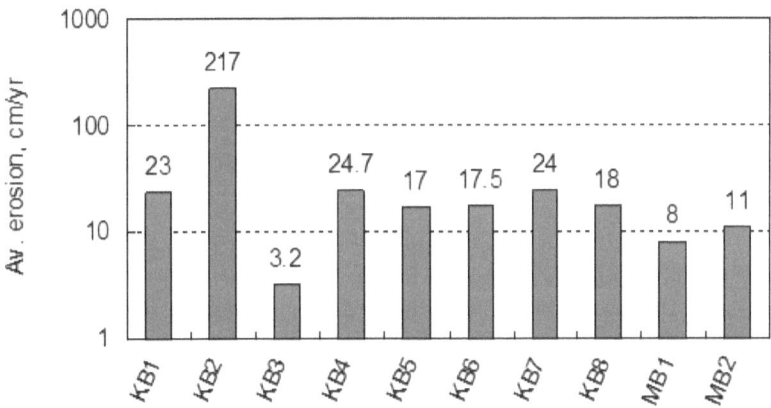

Figure 9.4 Average annual rates of bank erosion (May, 2002-Aug, 2004).

The temporal change pattern in erosion rates was examined if it can be explained by the change in rainfall pattern. However, erosion response to the rainfall was widely different for each bank sites. It shows that the temporal and spatial variation can not be described only by the variation in the rainfall input which further suggests that other controlling factors must be considered.

Controlling factors of bank erosion

From the results of erosion rates alone, it is difficult to define the contribution from individual control factor. However, from the results combined with field evidence, it can be asserted that bank composition appeared to have bigger effect on erosion at a catchment scale. It appears that the rate of erosion is closely linked with material composition of the bank.

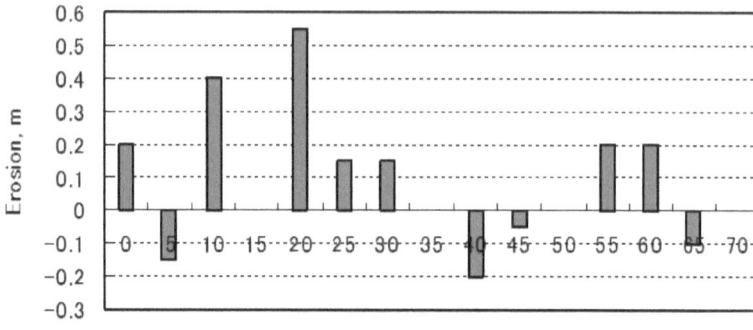

Figure 9.5 Erosion amounts indicated by individual pins in MB1.

At KB2, gravel and boulder mixed soil which is generally highly erodible was exposed up to the high flood level. In contrast, in the rest of the banks high flood generally occurred up to the layers of fine grained soil mostly clay and silt. At KB3 relatively less amount of erosion was possibly due to the higher consolidation of the fine-grained layers within the bank.

In order to assess the erosion resistance of the soil layers, field measurement was undertaken for shear stress at each bank site. The measurement was conducted at the bottom layer of each bank, which is mostly composed of clay and silt. Shear strength was measured by using a vane shear tester (Model STCL-4, QCQA media LLC, US) and a penetrometer (Model STCL-3). The measurements were carried out during the rainy season during which the banks are likely to be in fully saturated condition.

The shear strength of fine-grained bottom soil layer varies from 0.26 to 0.88 Kg/cm^2 (Table 9.2). Stream banks KB2 and KB6 had the lowest strength and KB3 had a highest strength. However, penetrometer test results indicated not much significant variation across the banks. Difference in shear strength at the banks KB2 and KB3 implies different degrees of soil consolidation prevailing within the banks. Since the two banks were found to have maximum and minimum erosion respectively, the test results can partially explain the inverse relationship between the erosion and bank strength

From the perspective of bank vegetation, no isolated effect can be identified from the measurements. But on the basis of field observation, higher erosion of KB2 may also be attributed to the presence of shrubs which are generally not effective during high flows (Thorne, 1990). Again, this assumption can not be valid for other banks with shrubs which have shown relatively lower erosion rates.

In the same way, the results indicate that factors such as bank height and bank slope have shown no coherent relationship to erosion or it may also be possible that their effects are outweighed by other factors.

Table 9.2 Shear strength measurement at the study stream banks. Bank strength are shown as an average±SD. Sample size=5 for each bank.

Bank	Shear strength Kg/cm^2	Penetrometer Kg/cm^2
KB1	0.35±0.65	3.80±0.55
KB2	0.26±0.48	3.90±0.55
KB3	0.88±1.00	4.30±0.27
KB4	0.35±0.42	4.20±0.27
KB5	0.38±0.50	4.10±0.22
KB6	0.26±0.37	3.80±0.45
KB7	0.29±0.61	4.20±0.27
KB8	0.40±0.22	4.10±0.22
MB1	0.38±0.35	4.30±0.27
MB2	0.40±0.22	3.90±0.55

Erosion processes

Bank erosion takes place mainly in the period of prolonged rainfall by a variety of reasons (Box 9.4). The process of erosion is complex and widely varied as mentioned earlier. The streams change their course frequently and the processes are further complicated by the recurrent changes of flow course within the channel itself. As a result, formation of small island bars is also frequent which can alter the flow direction bringing changes in erosion pattern. Here some of the typical erosion processes are described which can clearly be visible in the stream channels.

> **Box 9.4 A stable bank can be transformed into an unstable bank during period of prolonged rainfall through (Simon et al., 2000):**
>
> - increase in soil bulk weight,
> - decrease or complete loss of matric suction and therefore apparent cohesion
> - generation of positive pore-water pressures, and therefore, reduction or loss of frictional strength, entrainment of in situ and failed material at the bank toe and
> - loss of confining pressure during recession of stream flow hydrographs.

Bank failure mechanism

Bank failure mechanisms depend on many factors; however, predominant ones are the bank geometry and type of material. Failure mechanisms can broadly be divided into four types (USDA, 2003): 1. Rotational failure , 2. Planar failure, 3. Cantilever failure and 4. Piping failure (Fig. 9.6). There is a clear contrast in failure mechanics between non-cohesive and cohesive materials. Deep-seated failure occurs in cohesive materials with a block of disturbed material sliding into the channel along a curved failure surface. Steep banks, on the other hand, composed of non-cohesive materials fail along almost planar surfaces.

At the stratified composite banks of Siwalik streams, cantilevered or overhanging banks are generated when erosion of an erodible layer leads to undermining of overlaying, erosion-resistant layer (Fig.9.7). The strength of cantilever blocks is significantly increased by root reinforcement due to riparian vegetation and decreased by tension cracks.

At the banks composed of cohesive materials - clay and silt, with intermittent layer of erodible sand, piping failures can be observed. The piping undermines the overlaying bank materials, which then collapse. In the study area, this type of failure is mainly noticeable in the tall banks. Examples of such failures are presented in Fig. 9.8.

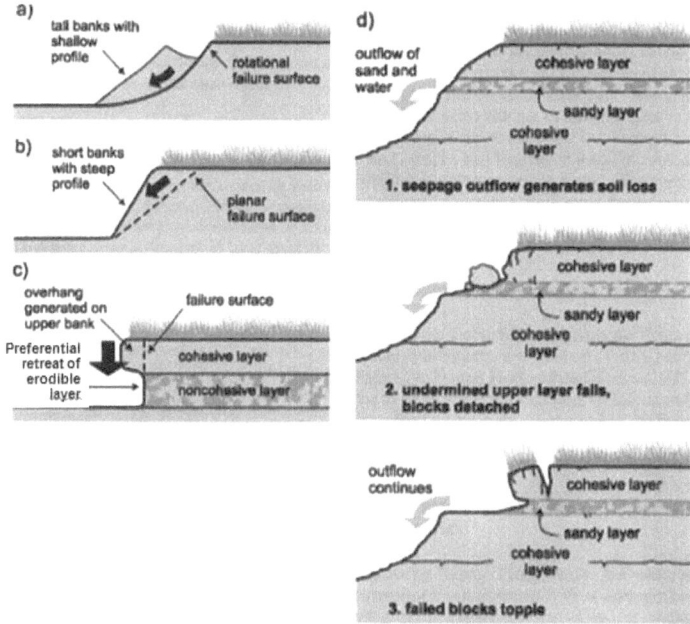

Figure 9.6 Illustration of different types of failure mechanisms in stream banks (USDA, 2003).

Figure 9.7 A composite bank of Khajuri Khola undergoing cantilever failure.

Figure 9.8 Piping failure at the tall banks composed of silt and clay layers.

Stream bank - vegetation interaction

As outlined earlier, vegetation can have both positive and negative effects on bank stability. In the study streams, both of these effects are evident. Positive effects can be witnessed in tall banks particularly in the reach where stream width is generally wider. In such banks, debris talus formed by the fallen bank materials protects the bank-toe from further erosion. Many studies have outlined this effect of bank accretion as the failed bank materials do have some apparent cohesion (e.g. Wood et al., 2001, Lawler et al., 1997). The banks can further be strengthened, if the talus material is favourable to vegetation growth. Fig. 9.9 shows an example of this effect in one of the study streams. It is important to note here that the process of vegetation recovery around the talus deposit can be a basis for planning of channel restoration works. For this, design of protection works must be based on the factors that influence channel instability in order to protect the failed materials until full recovery of vegetation is gained.

The effect of tree roots on bank erosion was found less significant especially in the banks composed of stratified layers. There may be two reasons: first, most of the bank vegetation consists of young Sal (Shorea Robusta) trees which are yet to develop the rooting network in full capacity. So, the roots are of shallower in depth. Second, the topsoil matrix is shallow in depth so that root development is usually restricted by the layers composed of coarse gravel and boulders. In contrast, relatively dense root network was observed in the banks composed of fine grained materials such as fine silt and clay. However, proportion of such banks is less compared to the stratified banks. On the negative side, isolated matured trees have promoted instability of stream banks.

In the study stream the large tree trunks which are not in well-colonized position promote local pockets of erosion leading to possible destabilization of themselves. It is evident by frequent fall down of tall trees along the stream banks (Fig.9.10). As outlined earlier, the effect of vegetation depends on many factors, among them type, density, position and spacing are the important ones. Promotion of near-bank vegetation is thus important to maintain their proper spacing.

Figure 9.9 Natural process of bank recovery through vegetation growth in Khajuri stream.

Channel change through cutoff

As also evident from the analysis of channel course change, the main flow can change its direction through cutoff bringing abrupt changes channel morphology and erosion processes. The bank site KB3, for example, did not suffer any erosion in 2004 because of change in main flow direction through cutoff. However, a rigorous scoring was observed on the newly formed cutoff channel. Investigation on the occurrence of cutoff and its effects on bank erosion is beyond the scope of the present study, however, it is found that this process is prevalent in many other stream channels indicating a rapid change in channel morphology.

Figure 9.10 Bank instability caused by toppling of a tall tree in Khajuri stream.

Bank erosion hazard mapping

Background

The main purpose of bank erosion hazard mapping is to evaluate the erosion potential of stream banks by assessing the governing parameters in the scale of relative hazard rating. In light of the complexities of the bank erosion processes, empirically derived rating methods can be helpful to integrate the governing parameters.

In this study, hazard-mapping method developed by Rosgen (2001) has been used by making appropriate adjustments to fit in the context of study area. The potentiality of the adopted method has been evaluated independently on the basis of stream planform change during the last four decades (as presented in Chapter 4).

Method

Two indices were developed by Rosgen (2001): Bank erosion hazard index (BEHI) and near-bank stress (NBS). Field measured variables were converted into risk rating of 1 to 10 (10 being the highest level of risk). The total points obtained from the conversion of the measured bank variables to risk rating indicate different scale of hazards. The risk ratings from 1 to 10 indicate corresponding adjective values of risk of very low, low, moderate, high, very high and extreme potential erodibility (Table 9.3).

Rosgen (2001) adopted five key field parameters as bank erodibility variables namely: ratio of bank height and bankfull height, ratio of rooting depth and bank height, root density, slope steepness and percent of surface area protected. In addition, risk rating has been defined by velocity gradient and ratio of near-bank stress and shear stress. By combining the two indices, he tested the method by taking independent data sets from two varied hydro-physiographic regions.

Rosgen (2001) further notes that adjustment can be made for the bank materials assigning 5 to 10 points depending on the material types. Similarly 5 to 10 points can be assigned for the stratification depending on the number and position of layers.

Application

In the context of study area, some field variables have been modified and some new ones added. Appropriate index values are assigned based on the relative contribution of controlling parameters which is rather empirical.

Since the majority of the bank profiles are nearly vertical, the parameter-bank angle was discarded. Vegetation effect is represented only by root depth in relation to the bankfull height. The parameter-surface protection is discarded as virtually no large-scale protection work exists. Velocity gradient and near-bank stress are taken into account in view of the steep slope gradient of the streams. For this, two variables were found significant: stream gradient and angle of bend.

It was deemed important to incorporate the bank material stratification as a new variable in the context of composite banks. While the banks are highly stratified, the material layer that comes into contact of floodwater directly deserves most important parameter for erosion. Hence, six category of bank layer (usually bottommost) are distinguished, ranging from cohesive clay/sand to non-cohesive gravel/boulder. Bankfull heights were measured at regular interval of the streams. Where the height was not clear, bankfull level was judged by means of watermarks on the banks Depth of exposed roots was directly measured in the field. Slope gradient of the streams was computed undertaking leveling survey. Bend angles of the stream were measured on the topographic map. The bank variables and hazard ratings are presented in Table 9.4. Fig. 9.11 shows the hazard map of Khajuri stream based on the total index computed for each hazard class. Computation of hazard rating for the Khajuri stream is shown in Table 9.5.

Table 9.3 Streambank characteristics used to develop bank erosion hazard index (BEHI) Rosgen (2001).

Risk level		Bank H/ Bankfull H	Root D/ Bank H	Root density %	Bank angle Degrees	Surface protection %	Totals
Very low	Value	1.0-1.1	1.0-0.9	100-80	0-20	100-80	
	Index	1.0-1.9	1.0-1.9	1.0-1.9	1.0-1.9	1.0-1.9	5.0-9.5
Low	Value	1.11-1.19	0.89-.50	79-55	21-60	79-55	
	Index	2.0-3.9	2.0-3.9	2.0-3.9	2.0-3.9	2.0-3.9	10.0-19.5
Moderate	Value	1.2-1.5	0.49-.30	54-30	61-80	54-30	
	Index	4.0-5.9	4.0-5.9	4.0-5.9	4.0-5.9	4.0-5.9	20.0-29.5
High	Value	1.6-2.0	0.29-0.15	29-15	81-90	29-15	
	Index	6.0-7.9	6.0-7.9	6.0-7.9	6.0-7.9	6.0-7.9	30.0-39.5
Very high	Value	2.1-2.8	0.14-0.05	14-5	91-119	14-10	
	Index	8.0-9.0	8.0-9.0	8.0-9.0	8.0-9.0	8.0-9.0	40.0-45.0
Extreme	Value	>2.8	<0.05	<5	>119	<10	
	Index	10	10	10	10	10	46.0-50.0

Table 9.4 Streambank characteristics used to develop bank erosion hazard index (BEHI) of Khajuri stream.

Risk level		Bank H/ Bankfull H	Root D/ Bank H	*Material	Slope %	Bend Degrees	Totals
Very low	Value	1.0-1.1	1.0-0.9	1	<0.40	166-180	
	Index	1.0-1.9	1.0-1.9	1	1.0-1.9	1.9-1.0	5.0-9.5
Low	Value	1.11-1.19	0.89-.50	2	0.41-0.80	151-165	
	Index	2.0-3.9	2.0-3.9	3	2.0-3.9	3.9-2.0	10.0-19.5
Moderate	Value	1.2-1.5	0.49-.30	3	0.81-1.20	136-150	
	Index	4.0-5.9	4.0-5.9	4	4.0-5.9	5.9-4.0	20.0-29.5
High	Value	1.6-2.0	0.29-0.15	4	1.21-1.60	121-135	
	Index	6.0-7.9	6.0-7.9	6	6.0-7.9	7.9-6.0	30.0-39.5
Very high	Value	2.1-2.8	0.14-0.05	5	1.61-2.00	106-120	
	Index	8.0-9.0	8.0-9.0	8	8.0-9.0	9.0-8.0	40.0-45.0
Extreme	Value	>2.8	<0.05	6	>2.00	<105	
	Index	10	10	10	10	10	46.0-50.0

*Predominant material up to the bankfull height (1-Bed rock, 2- Clay, 3- Silt, 4- Sand, 5-Gravel and 6-Gravel/boulder)

Evaluation

Qualitative assessment can be done in a variety of ways in order to examine how best the technique can yield the results. Winterbottom and Gilvear (2000), for example, evaluated the bank erosion probability model developed by Graf (1984) by using GIS-assisted mapping method of planform change. Similar attempt has been made in the present study to evaluate the technique based on the stream response to the morphological changes in stream planform. The fundamental idea is that changes in stream planform indicate, in some form, width adjustment which in turn depends on bank erosion (or advance). Using the information derived from Chapter 4, a scale of planform change is constructed using five important planform change patterns (Table 9.5).

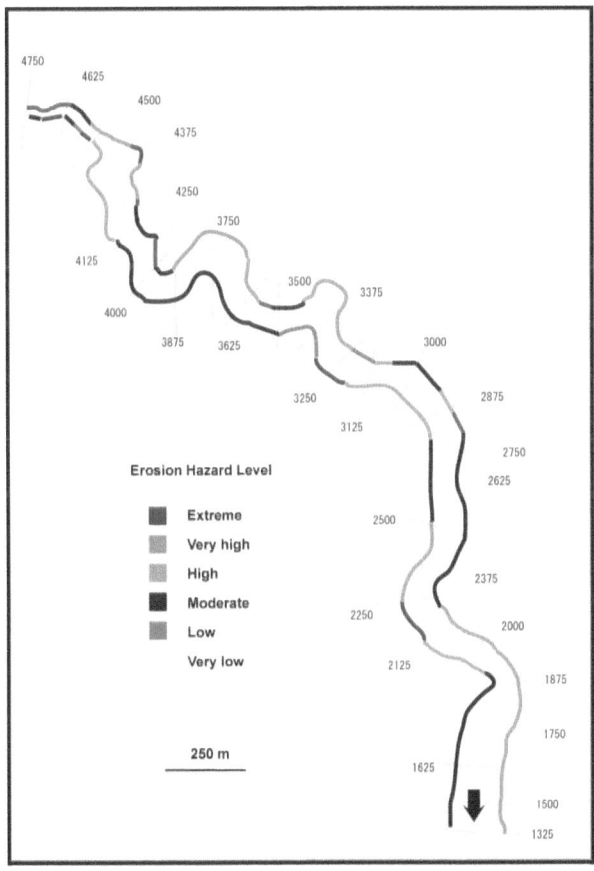

Figure 9.11 Bank erosion hazard map (BEHM) of Khajuri stream.

Based on the risk class defined in Table 9.5, channel adjustment map was prepared. It is important to note that the planform change pattern was assessed qualitatively where some logical judgments were also applied. For example, very minor changes in bank lines were discarded as they could be the result of residual errors that arise from rectification of aerial photograph or tracing bank lines.

Table 9.5 Planform change categories.

Planform change class	Description of planform change (1964-1992)	Implication to erosion
1 (lowest)	Narrowing	Very low (no erosion)
2	Stable reach with virtually no change in planform	Low erosion
3	Reach of gradual widening	Moderate erosion
4	Active widening	High erosion
5 (highest)	Course shifting	Very high erosion

The two independent maps were then compared by overlaying them which is shown in Fig. 9.12. It indicates that in a majority of the sections, the two rating systems match well. However, in some locations like in sharp bends the rating classes differ. The reason can be explained by the fact that high banks generally exist in the sharp bends which exhibit different behaviour in terms of bank retreat. Virtually no bank retreat can be detected from the aerial photographs in such bends while the erosion on the toe is generally active. From this fact, the evaluation technique needs to be further revised and considered for the erosional activities in tall banks in the sharp bends.

If the two rating systems are compared by the percentage of each class, it would indicate quantitatively how well the two systems match. The comparison is shown in Fig. 9.13. The figure implies that main classes of risk are moderate and high in both methods. Maximum difference in the moderate and high class is about 25%. Hence, considering this discrepancy as reasonably fair, it can be asserted that the two methods exhibit a good match and the developed method can be a potential for the evaluation of bank erosion hazard in the Siwalik streams

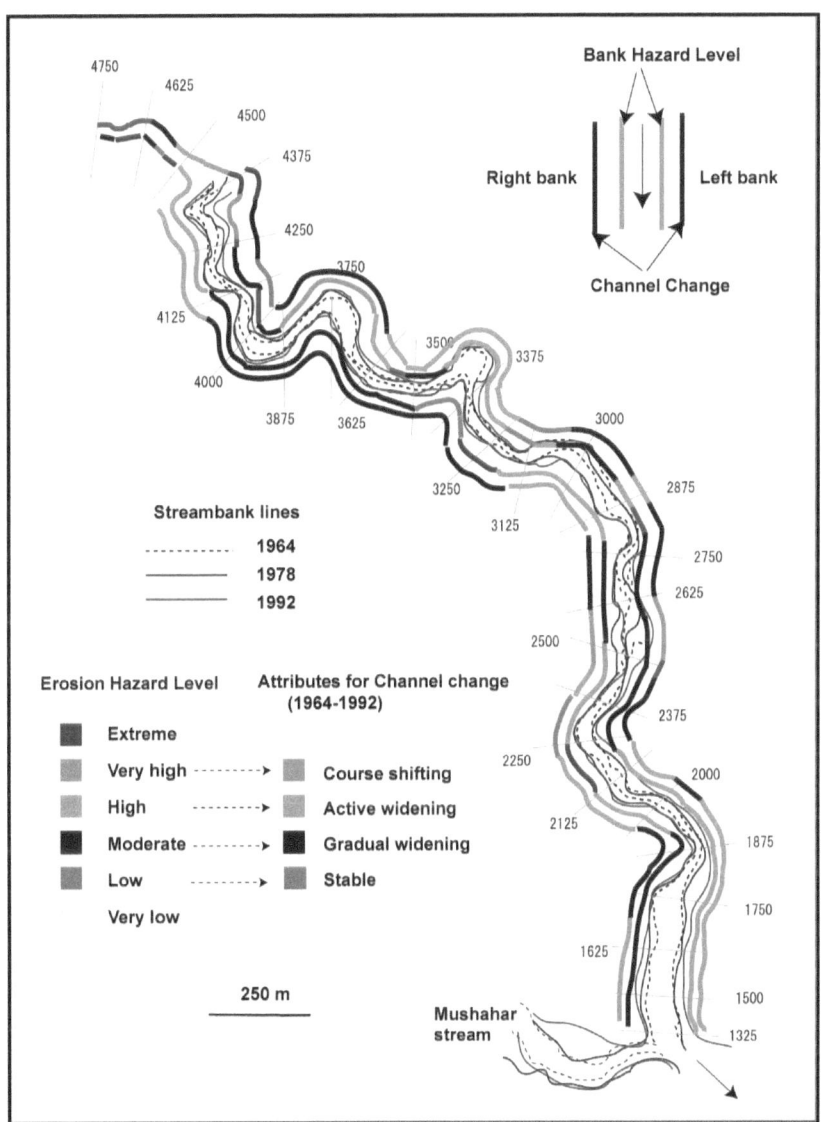

Fig. 9.12 Overlay of stream planform and bank erosion hazard map.

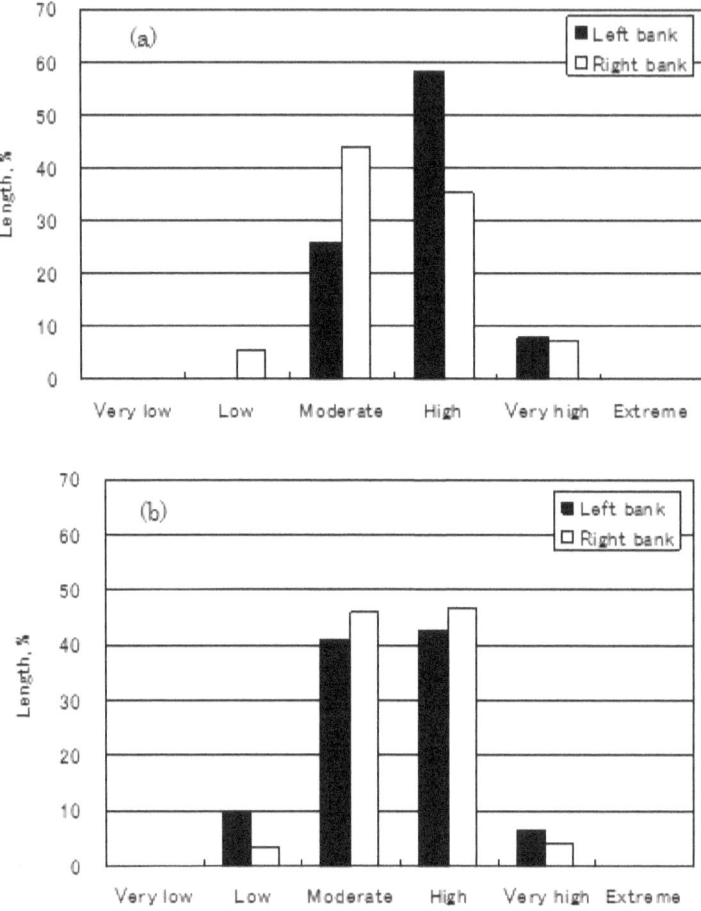

Figure 9.13 Comparison of hazard classes based on the length of bank, (a) Bank erosion hazard rating, (b) Channel adjustment hazard rating

Summary

Following main points can be summarised:

- Bank erosion is a complex process with many controlling factors, the isolated effects of which is difficult to quantify.
- Bank erosion rates varied widely both in space and time. Erosion rates were found to be from few centimeters to as high as 2 m/year. However, in the forested area, average erosion was 12 cm/year.
- Main processes of bank erosion can be categorized as subaerial, fluvial and mass failure. In the stratified composite banks of Siwalik

streams, cantilevered or overhanging banks are generated when erosion of an erodible layer leads to undermining of overlaying, erosion-resistant layer.
- Vegetation can have both positive and negative effects on bank stability, although it generally inhibits fluid entrainment. In the study area, no distinct effect of vegetation could be quantified, but some positive effects were observed where vegetation recovery occurred in failed material deposits. In contrast, fallen trees were also evident in the streams. It implies that the spacing or density of vegetation should be sufficient to protect the banks.
- A method of evaluation for bank erosion hazard has been introduced based on field-measurement of controlling variables. In this method, bank erosion hazard index (BEHI) and near-bank stress (NBS) variables have been combined and transformed into the scale of risk rating. The method is tested independently by using channel adjustment hazard map based on aerial photo overlay of banklines in different time periods. Comparison results are found satisfactory, indicating the potential of the method for the bank erosion hazard assessment of Siwalik streams.

10
Estimation of sediment at a catchment scale

Background

The purpose of this chapter is not only to determine the absolute values of the sediment production rates, but also to ascertain the relative contribution of different types of sediment sources at a catchment scale. The total sediment production has been estimated by extrapolating the sediment production rates from a plot scale to the catchment scale. It is noteworthy to mention that the extrapolation may have many limitations. The accuracy of estimation obviously depends on how well the plot data represent the whole system at the catchment scale. Estimation of the catchment-wide sediment production has been undertaken taking into account of the four major sediment sources- surface, gully, landslide slope and streambank erosion as outlined in the previous chapters.

Surface erosion

The sediment produced from surface erosion has been estimated based on landuse because no significant difference in erosion rates was observed in relation to the slope gradient. In the Siwalik Hills in Haryana-Panjab of India, Kukal et al., 1991 found that vegetation cover was a more vital factor than slope steepness and length in determining actual erosion.

Since the erosion rates from the forest and shrub land do not differ much, a common average erosion rate of 1.0 ± 1.0 mm per year has been considered. Distinction has been made between vegetated and non-vegetative (principally agricultural) land. Since erosion from agricultural land was not monitored, an estimated value of 2 mm per year has been considered for gently sloping rain-fed terrace land. A similar rate is reported by Merz, 2004 in the Middle Hills where down-sloped agricultural land suffered from erosion ranging from 1 to 2 mm every year. Table 10.1 shows the computation of sediment volume based on these average rates. The table revels that about 65% sediment is generated from the forest and shrub land whilst rest 35% from agricultural land.

Table 10.1 Computation of average annual sediment production rate and total volume from surface erosion.

Land cover	Area		Av. Erosion		Total sediment vol.
	km²	m²	mm	t/ha	m³
Forest and shrub land	3.464	3464000	1.0 ±1.0	14±14	3464±3464
Agriculture	0.933	933000	2.0	28	1866
				Total	5330 ± 3464

Gully erosion

As there are three large active gullies within the study catchment, sediment produced from these gullies namely KG1, KG2 and KG3 has been summed up. From a gully, sediment is assumed to be generated from the retreat of headwall and erosion of sidewalls. Since erosion rates from headwall and sidewall differed much over the monitoring period, an average value has been considered for computation (Table 10.2). It indicates that gully KG1 is the most active gully producing about 50% of sediment produced from all gullies together.

Table 10.2 Computation of annual average sediment volume from gully erosion.

Gully	Area	Av. sediment vol.	Total sediment vol.	Average rate	
	ha	m³	m³	m³/ha	t/ha
KG1	0.60	1766 ± 273			
KG2	0.36	751 ± 39	4169 ± 440	2265 ± 239	3770 ± 335
MG1	0.88	1652 ± 128			

Landslide slope erosion

Landslides have been mapped from aerial photography followed by field verification. Total area (plan area) is calculated by summing up area from individual landslides. From the erosion measurement of three representative landslides, average volume of sediment produced from unit area is computed, which is then used to estimate sediment production from all landslides. It is important to note that, in estimating the sediment production from the landslide slopes, it is assumed that the erosion takes place in parallel throughout the landslide slope in a uniform fashion. This assumption is based on the hypothesis of slope evolution put forward by Young (1972). According to this hypothesis, a slope may be developed by the equal retreat (such as by erosion) throughout its slope, and the free-face (like the head scarp of landslide) also retreats in parallel with the slope surface. This evolution generally belongs to the long timescale during which the free-face retreats by toppling (Fig. 10.1). Computation of the sediment volume is shown in Table 10.3.

Obviously, the assumption can affect the accuracy of estimation since the erosion measurement on the landslide slope surfaces belong to short time period. But it is asserted that the erosion takes place actively also on the head scarp slope of the landslides as the topography is steep and materials are weak, thus enabling parallel retreat of the sliding slope and head scarp slope.

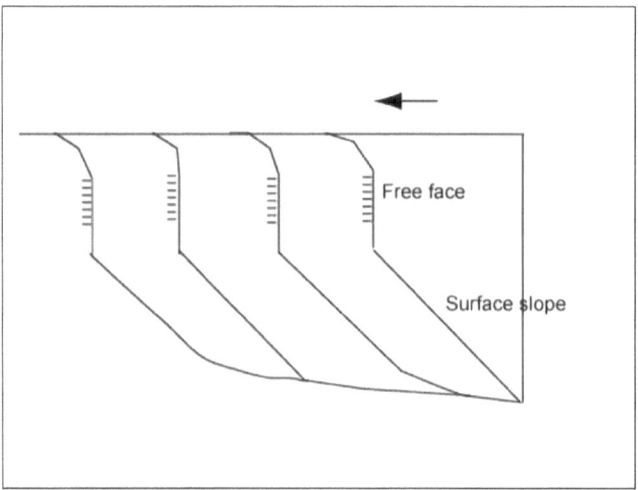

Figure 10.1 Principle of parallel retreat (Young, 1972).

Table 10.3 Computation of sediment volume from landslides.

Landslide	Slope area, m²	Slope, degree	Plan area, m²	Av. erosion cm	Volume m³	Volume/ Plan area m³/m²
L1	1200	50	768	6 ± 3	72 ± 36	0.10 ± 0.04
L2	980	45	686	6 ± 1	58.8 ± 9.8	0.09 ± 0.01
L3	645	40	490	5 ± 1	32.3 ± 6.4	0.07 ± 0.01

Total landslide area		Av. sediment volume/ area.	Total sediment vol.	Average rate from landslide area
km²	m²	m³/m²	m³	t/ha
0.15	150000	0.08 ± 0.02	12000 ± 3000	1120 ± 280

The table shows that the landslides with different slope area produce sediment differently; however, ratio of sediment volume and plan area varies closely between 0.07 and 0.10. This is mainly due to the fact that the landslides exhibit quite similar characteristics in terms of surface erosion. Also, the slope inclination of the landslides does not vary widely.

The sediment production from the landslide areas within the catchment would be 12000 ± 3000 m³. This means the annual average rate of sediment is 1120 ± 280 t/ha.

Bank erosion

In order to assess the significance of bank erosion in the catchment scale, bank erosion rates derived from the field measurements were extrapolated neglecting the spatial variation in erosion rates. Instead; average value of erosion rate was considered in the estimation of sediment production.

Three types of bank categories were defined: stable bank and unstable bank. Stable banks, in this context, are defined as the banks which show no distinct cut-slope. They are generally low-height banks usually less than 1m, which may have developed fair vegetation cover on the face. Unstable banks are subdivided into two types- short and tall cut bank, as erosion rates and processes were apparently different in these two types of bank. Tall banks (usually more than 7m in height) were further divided into two types in relation to the erosion activity indicated by vegetation: active and semi-active. Active tall banks have little or no vegetation cover; on the other hand semi-active tall banks develop sparse vegetation on the bank face and talus deposit.

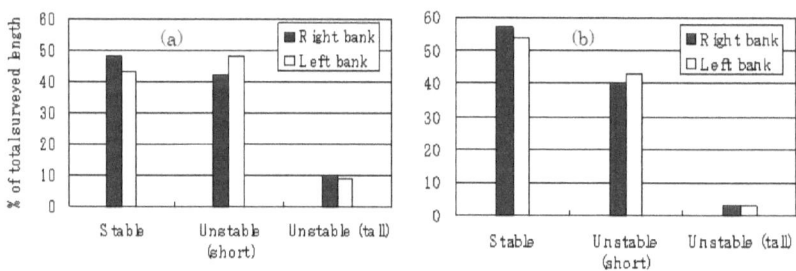

Figure 10.2 Types of banks in (a) Khajuri, stream, total length 4.7km and (b) Mushahar stream, total surveyed length 3 km.

Table 10.4 Computation of annual average sediment volume from bank erosion.

Bank type	#Total area m²	Av. Erosion cm	Av. volume m³	Total sediment vol. m³	* t/km
Short bank	5719	16 ± 5	915 ± 286		
Tall bank				1730 ± 810	367 ± 172
-Active	5131	12 ± 8	616 ± 410		(524 ± 245 t/km²)
-Semi-active	2839	7 ± 4	199 ± 114		

\# In Khajuri stream, surveyed lengths in Khajuri and Mushahar streams were 3.3 and 3 km respectively. About 45% of the total length in Khajuri and 55 % of that in Mushahar belong to stable banks which are not considered in the sediment estimation.

* Total length = 6.6 km (including both streams).

Field measurements were undertaken for length and height of each type of banks along the two streams- Khajuri and Mushahar in order to estimate total bank area undergoing erosion. Stable banks include about 45 and 55% of the total surveyed length in the two streams respectively (Fig. 10.2). The figures for the unstable short banks are 55 and 42%. The occurrence of unstable tall banks is less in both streams. It is important, however, to note that there is wide spatial variation on the occurrence of the bank types along the two sides of the streams.

Sediment production has been estimated by multiplying the bank area in each category and respective average erosion rates, a similar method adopted by Stutt (1997). For short bank, average erosion rate (16 cm) has been considered excluding the extravagant erosion at bank KB2, while average measured values of 12 and 7 cm are taken for the active and semi-active type of tall banks. The estimation of sediment volume is presented in Table 10.4.

Catchment-wide sediment production

When summing up the sediment amounts from each type of sediment source as estimated above, it yields total sediment produced from the catchment. The summary of the erosion rates and sediment production is given in Table 10.5. Relative contribution of each type of sediment source to the total sediment from the catchment is shown in Fig. 10.3. It indicates that majority of the sediment is contributed by landslides/slope failures (48%). Relative contributions from surface erosion and gully erosion stand at 22% and 18% respectively. Contribution from the bank erosion remains at 12%.

Table 10.5 Summary of erosion rates and annual average sediment production.

Sources	Average annual rate	Total sediment volume in catchment scale, m^3/yr
Surface erosion	14 ± 14 t/ha/yr (forest), 28 t/ha/yr (agricultural land)	5330 ± 3464
Gully erosion	3770 ± 335 t/ha/yr	4169 ± 440
Landslide slope erosion	1120 ± 280 t/ha/yr	12000 ± 3000
Bank erosion	367 ± 172 t/km	1730 ± 810

Table 10.6 summarises the annual average rate of sediment production from the whole catchment. It indicates that on average the study catchment (4.62 km^2) generates approximately 23230 m^3 of sediment annually, which is equivalent to 7039 tonnes/km^2 considering the bulk density of composite sediment as 1.4 tonnes/m^3. Also, when it is converted to the erosion depth, about 5 mm of soil loss occurs from the catchment.

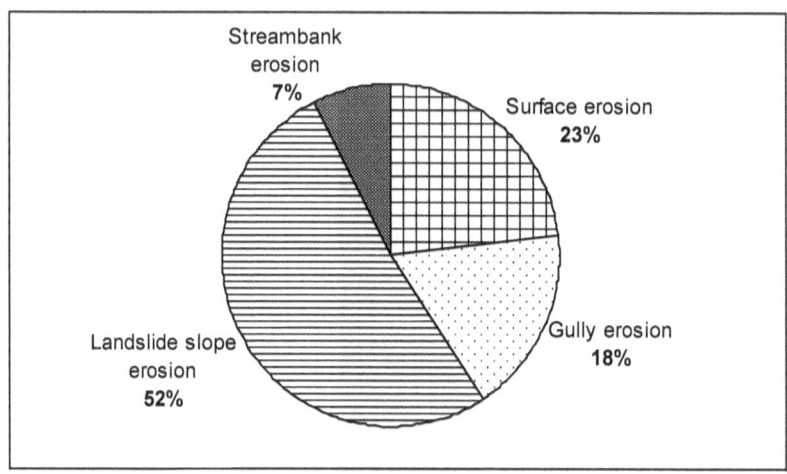

Figure 10.3 Relative contributions of sediment sources in the Khajuri catchment in terms of total sediment volume.

Table 10.6 Summary of annual average rate of sediment production from catchment

Total sediment volume	23229	m^3
Total sediment weight	32521	t
Total catchment area	4.62	km^2
Rate of sediment volume	5028	m^3/km^2
Rate of sediment weight	7039	t/km^2
Rate of erosion depth	5.0	mm

Erosion hazard map

By combining the erosion rates and processes at the catchment scale, an erosion hazard map has been developed (Fig. 10.4). This map shows the location and distribution of major sediment sources in the catchment, as well as the average annual rates of erosion obtained from the field measurement. It is important to note that the erosion rates form surface, landslide slope and gully are expressed as sediment weight produced per unit plan area of these features per unit time, such as t/ha/year. However, rate of bank erosion is taken as the erosion depth perpendicular to the longitudinal profile of the bank, often expressed as distance per unit time, such as m/year. For this it is necessary to multiply by the exposed area of the bank and unit weight of the bank material.

It is asserted that this type of map is of high relevance and useful in estimating the sediment yield from an unguaged catchment in the Siwalik region. It is also helpful for comparing the degradation status of the catchments. Therefore, the catchment-scale rate and process map can be an important tool for planning and designing of conservation works.

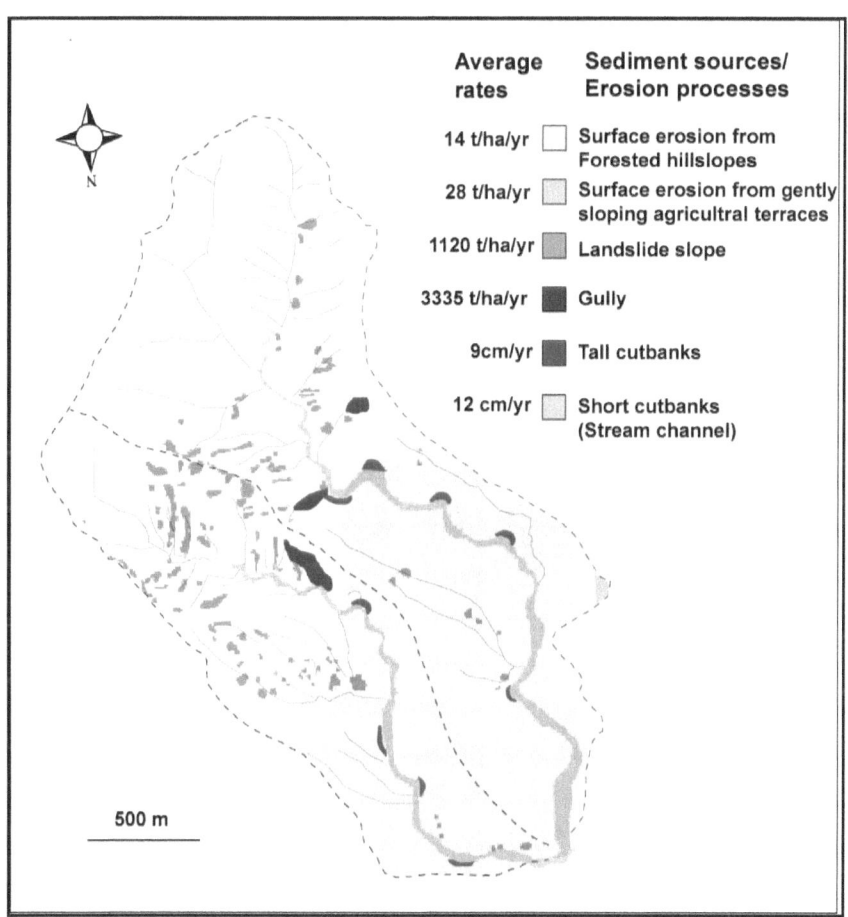

Figure 10.4 Erosion hazard map of Khajuri watershed.

Sediment yield

Fig. 10.5 shows that suspended sediment yield generally increases with the increase in rainfall size, however due to very limited number of measured events effect of rainfall intensity could not be ascertained. It is noteworthy to mention that the major weakness of sediment yield estimation is the estimation of bedload, which could not be measured in the stream.

Total sediment production and sediment yield as computed above seem to be quite in balance leaving no room for bedload which can be contradictory. There could be many reasons to explain this difference. Erosion and mass wasting never occur in a uniform fashion as supposed, nor does the sediment delivery pattern follow the consistent trend. However, the comparison does indicate a minor difference in sediment production and yield, which is quite plausible in case of the catchments of smaller spatial scale.

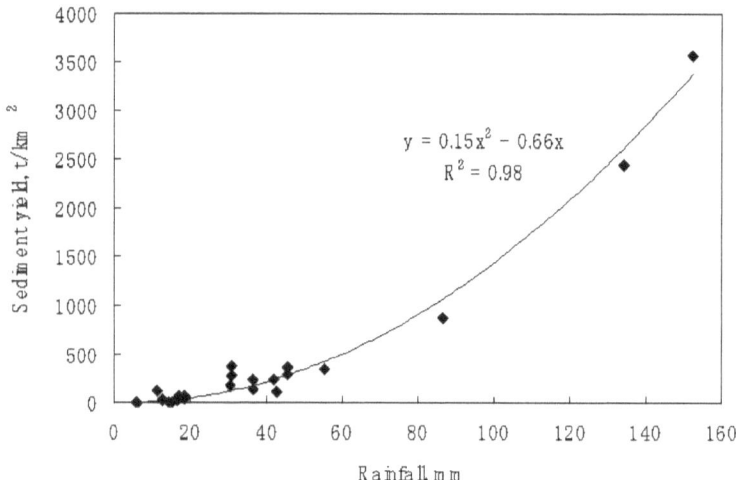

Figure 10.5 Sediment yield from different rainfall events.

Not all the material that is eroded will leave the catchment, as quite a lot of the eroded material may bee re-deposited within the catchment itself. To account for this interaction of erosive and depository processes, the sediment delivery ratio (SDR) must be calculated. The SDR represents the fraction of the material eroded from a particular catchment which reaches the outlet of the given catchment and where the sediment is measured. As the number of storage opportunities tends to decrease with catchment size, the SDR increases markedly for small drainage basins (Walling, 1983). Field observation also suggests that no large sediment deposition features exist in the catchment. This indicates that SDR can tentatively be taken as unity for the steep channels like the study stream.

Comparison to other studies

While acknowledging the disparities mainly in the scale and management strategies, it is attempted to compare the results obtained from this study with that from other studies. Comparing the results, however, is a difficult task because studies dealing with sediment production in Siwalik Hills are very limited. Some of them have reported approximate erosion rates from a variety of landuse systems, which are summarized in Table 10.7. The table reveals that erosion rate varies widely in relation to the land use. For forest, for example, the rate varies from 780 to 4000 t/km^2/yr depending on the degree of degradation. Also, it is apparent that soil loss from severely degraded gullied land is excessively high. The rates derived form this study is found to be well within the large variability of soil loss from the past studies. For forest and shrub land, annual average erosion of 1 mm (1400 t/km^2) can be considered plausible in the context of increasing grazing pressure. While taking account of sediment production rate from the whole catchment (7400 t/km^2), it can be regarded as a severely degraded catchment.

Table 10.7 Summary of erosion rates in Siwalik Hill region.

Description	Erosion rate t/km²/yr	Reference
Foothills of eastern Siwalik of Nepal, ranging from forest to grazing	780-3680	Chatra, 1976
Siwalik Hills of far western Nepal		
Degraded forest	2000	Laban, 1978
Degraded forest, gullied land	4000	Laban, 1978
Ratu Khola watershed, Siwalik Hills	5000	Honda et al., 1996
Surkhet, west Nepal, severely degraded, heavily grazed forest, on intensively gully badlands	20000	Sakya in Laban, 1978
Nurpur Siwalik Hills of India, heavily grazed forest	2200	Ghosh and Subba Rao, 1979
Siwalik Hills of Panjab-Haryana, India	2000-8000	Singh et al., 1992
Siwalik Hills of Chandigarh, small basins, cultivated terrace lands	250	Mittal et al., 1984
Rajpur, Deharadun of India, Shorea forest, typically grazed Siwalik headwater catchment	20	Subba Rao et al., 1973

Assessing degradation status of the catchment

LRMP (1986) has outlined the criteria used for the watershed condition classification. Five types of watershed condition are defined-class 1 to 5 respectively from excellent to very poor. The excellent class refers to the nearly undisturbed condition. While poor class refer to the occurrence of accelerated erosion.

The governing factors are percentage of ground cover, existence of rill, gully erosion and landslides, along with stream channel condition defined by extent of bank erosion and percentage cover of bank vegetation (Table 10.8). According to this classification, 37% of the area is under the excellent condition while 28 and 33% of the area can be classified as good and fair respectively. Only 2% of the area is classified as very poor condition.

Table 10.8 Evaluation criteria for watershed condition (LRMP, 1986). The bolded letters correspond to the study catchment.

Factor	Watershed condition class				
	Excellent	Good	Fair	Poor	Very poor
% of Ground cover	> 95	90-95	85-90	80-85	**<80**
Rill and gully erosion	Absent	Absent	**Present**	Present	Present
Landslides, ratio of accelerated to natural	<1.2	1.2-1.5	1.6-3.0	3.1-5.0	>5.0
Stream channel condition a. Bank erosion	None	Slight	**Moderate**	Weak braiding	Braiding, aggraded bedload
b. % of bank vegetation	>60	**30-60**	10-30	None	None

The main drawback of the classification system is its qualitative criteria, particularly about the occurrence of landslides. It is expressed as the ratio of accelerated to natural landslides. However, to define these two types of category is not easy, which is entirely based on the qualitative judgment. So, the classification system tends to sway largely towards the disturbance level by human activities. So, the question arises whether the landslides in the study catchment were natural or induced by human activities. In most of the cases, the field observation does not reveal the impact of human activities on the occurrence of landslides, meaning that the criteria belong to fair or good condition. When looking into the other criteria, such as presence of rill and gullies, and moderate level of bank erosion, the catchment tends to fall under the category of "fair" condition. The fair condition is defined by the existence of significant disturbances in the soil mantle and/or stream channels. A combination of education and structural remedies are required (LRMP, 1986).

Even though, the status of the catchment from the LRMP criteria seems to be fair, the catchment can be regarded as severely degraded based on the field measured sediment production rate of 7039 t /km^2 (5 mm of denudation rate). While comparing to other areas, the denudation rate of the study area belongs to the highest range in Nepal.

Summary

The following points can be summarised:
- Relative contribution of different sediment sources in the catchment sediment production has been found as: landslide slope erosion- 52%, surface erosion-23%, gully erosion-18% and streambank erosion- 7% on the basis of average sediment volume.
- Landslide slope erosion is the main process that is responsible for majority of the sediment production from the catchment, while gullies are the main point sources of sediment which have greatest sediment yield per unit area.
- Average annual catchment sediment production rate was about 7000 t/km^2 which is equivalent to the denudation rate of 5 mm/year.
- Based on LRMP's classification, the catchment tends to fall under the category of fair condition. However, from the viewpoint of sediment generation, the degradation status seems to be more severe.
- Catchment-scale distribution map of rate and processes of erosion is an important basis for planning and designing conservation works.

11

Soil water conservation and management policy

Background

In this chapter, it has been tried to bring up and highlight some conservation efforts with a focus to maintain and enhance the environmental integrity of the Siwalik Hills. Some examples are presented on the field-level applications of the conservation works, and at the end, management strategies have been suggested based upon the findings from the study.

Management policy

To conserve soil and water against degradation, we can broadly divide the programs/activities in the Siwalik Hills as follows:

- Micro-watershed management
- Integrated watershed management
- Degraded land rehabilitation
- River banks stabilisation

For long, Siwalik Hills ecosystem has been considered as an important component of Terai plain. Consequently, the earlier management strategies have focused on the preservation of the two regions collectively. Many of the programs based on this management strategy aim at sustainable livelihood of local people through the efficient utilisation of the available natural resources.

The most focused area is the forest management as the Siwalik Hills and its foothill (Bhawar) is covered by rainforest. Ultimate objective of the forest-related programs is to support poverty alleviation by encouraging participatory forest management and community-based initiatives. The Forest Act, 1993 and Forest Regulations, 1995 form the policy foundation for handing over national forests to local communities for conservation, management and utilization. The government has endorsed the policy provisions of Forestry Sector Master Plan and have given more thrust to people's participation in forest management practices, and in the programs, priority is set for the Siwalik Hills (MOFSC, 2002)

Some examples of soil water conservation schemes

The Biodiversity Sector Programme for Siwaliks and Terai (BISEP-ST) set forth by the government is another attempt which focuses on bio-diversity conservation and poverty alleviation through forestry sector development of Terai, Inner Terai and Siwaliks (DOFSC, 2002).

The government body- called Department of Soil Conservation and Watershed Management (DSCWM) is responsible for operating soil conservation and watershed management programmes with the following objectives (DSCWM, 2002):

- To contribute in meeting the people's basic needs for forest and food products by improving the productivity of the land through the conservation and management of watershed resources.
- To assist in maintaining ecological balance by reducing pressure from natural hazards such as floods and landslides.

In its strategy too, focus is given on the conservation activities in fragile landscape of the Siwalik Hills. From these facts on the policies and programs of the government, it is understood that emphasis is given for the degrading environment of the region; however, documentation on the achievements is very limited.

Based on the results from this study, two important issues can be highlighted regarding the management policy:

- Are the current policies/programs sufficient for addressing the water and sediment related problems in the Siwalik Hills?
- What types of concrete policy/programme are appropriate for dealing with the problems?

To analyse the first question, two field-based conservation schemes are discussed: Churiya Forestry Development Project (ChFDP) and Khajuri Model site of DPTC/JICA.

Churiya Forestry Development Project (ChFDP)

The ChFDP is a bilateral project of the Ministry of Forest and Soil Conservation supported by German Technical Cooperation (GTZ). Launched in 1992, the project has implemented many conservation and development programs mostly in the Siwalik (Churia) Hills extended within the three districts namelySiraha, Saptari and Udayapur in the eastern region. The project has identified the Siwalik Hills as an essential component of the Terai lowland, and has extended its program area to some parts of Terai.

The working theme of the project is mainly focused on the forest and the people (ChFDP, 2001). Alleviating poverty by creating alternative employment opportunities for off-forest income is one of its key objectives and the community forestry lies in the heart of the programme.

Figure 11.1 An earthen river embankment reinforced by shrub plantation (Baruwa Khola, near Gaighat, Udayapur, implemented by ChFDP).

Under its soil conservation program, various structural measures are tested and adopted such as riverbank protection works (Fig. 11.1), catchment ponds, fishponds and so on. Participation of the people was the principle concern of the scheme. The project claims the success of the community forestry programs with more than 200 users' group covering more than 30,000ha of forest in the project area (ChFDP, 2001). It has also reported that the recovery of the barren lands was possible by stabilising landslides and active gullies as a result of the programme.

As explained in Chapter 4, the forest cover in the southern hills of Trijuga River, which falls under the project area was found more or less unchanged during the period from 1992 to 2003 despite the deforestation trend in the past. Along with the geomorphological and topographical limitation for agricultural expansion, this positive result may partly be due to the success of its forestry programme. Moreover, the questionnaire survey conducted in the Khajuri catchment in 2003 also indicated that most of the beneficiaries expressed a good satisfaction over the achievement of the programme.

Khajuri catchment: a model site of DPTC/JICA

In 1998, Water Induced Disaster Prevention Technical Centre (DPTC)- a project run by the Government of Nepal with the assistance of Japan International Co-operation Agency (JICA) selected the study catchment of Khajuri Khola as a model site of erosion and flood control works. There were four types of conservation implemented at the site (Fig. 11.2):

(1) River control work such as gabion protected earthen embankment within the floodplain,
(2) Bank protection works with gabion walls,
(3) Gully protection works by gabion check dams and
(4) Plantation in the floodplain.

This project also encouraged the active participation of the local people in each activity, particularly in the plantation schemes. During the field works, the functioning of the applied measures was regularly monitored by means of repeated photography. A brief on the effectiveness evaluation and remedial ways are presented as follows:

River control works
The earthen embankments constructed on the floodplains of the Khajuri and its tributary stream Mushahar in 1998 were found functioning satisfactorily after 6 years since construction. The embankment height was decided to function at least for five years based on the high flood level corresponding to the rain burst of 1995. The aggradation rate of riverbed was estimated as 10 cm per year (Personal communication

with Higaki D.). However, no remarkable changes in bed level were noticed, which indicates that the stream continues to function as a sediment conveyance channel. It is also verified by the field questionnaire survey that almost all the respondents expressed much satisfaction over its functioning. With increased safety from flood, many new houses were built behind the embankments. In addition, a significant proportion of the floodplain is reclaimed to agricultural land.

- Local techniques using only soil and plantation works must be given priority in such works.
- The successful works must be up-scaled in other catchments.

Bank protection
Streambank protection works were found less effective clearly indicating the inappropriate selection of the streambanks. In many locations, the works were damaged by the stream current because of the improper orientation. Most actively eroded and vulnerable streambanks should be selected. The streambank erosion hazard map can be a useful tool for this purpose.

Figure 11.2 Different types of protection measures adopted in the experimental catchment of Kharjuri by DPTC/JICA (Gully control by gabion check dam, bank protection, earthen embankment and plantation behind the embankment respectively from top-left.

Gully control
Gully protection works by gabion check dams were found appropriate but insufficient. The high sediment debris from the active gully head still continued inducing lateral scouring across the check dams which triggered new shallow landslides.
- Protection work must be started from the gully head, as the study has indicated that it is the major sediment source than the gully channel.
- Series of small check dams on the drainage channel (of the gully head) combining vegetation could be effective.

Plantation
Plantation work was found successful but operation and management was not sufficient. On one hand, plantation helps to strengthen the embankment, on the other it supports for fulfilling the fuel demands in the community.

- Participatory approach is proven the most appropriate one for such conservation works.
- Strengthening of user groups is necessary by means of education, training and awareness schemes

Despite the limited patches of applied conservation works within the catchment, the scheme has demonstrated many important lessons dealing with the issues associated with the stream channels and erosional activities in the headwater reach. The above-mentioned activities were carried out after the preliminary assessment of catchment by means of instability mapping. Many active landslides and gullies were identified in the catchment, which form an important basis for this study.

Key issues about soil water conservation

Are the current policies/programs sufficient for minimizing water and sediment problems in the Siwalik Hills?
The policy of integrated management in the catchment scale, as advocated by the government in the poverty alleviation programs for tenth national plan seems an appropriate approach for the conservation of Siwalik Hill catchments. As the scale of the catchments is relatively smaller, the output of the collective actions is likely to be visible within a short temporal scale.

Keeping in view the positive results from the programmes like ChFDP and DPTC/JICA mentioned above, they can be taken as a good initiative towards safeguarding the fragile ecology of the region. However, more concrete action researches are utmost necessary since the comprehensive processes of ecological degradation and the interaction of the governing parameters are by and large unknown. The most important issue, also highlighted in this study, is the process of stream pattern change over time. The general assertion that change in land cover bring about changes in hydrology and morphology of the stream could not be proven in this study, which needs a more systematic study covering other similar catchments.

The policy and programmes adopted in other geo-physical environments, such as the Mountain regions or the Terai Plain can hardly be feasible in the Siwalik region because of the unique characteristics in terms of geomorphology, land cover and active tectonic activity. Research initiatives as regards the geo-environmental issues in the Siwalik Hills are largely lacking.

Despite the priority set in the conservation policies in the Siwalik Hills, not much have been done to transform the plans into action. The principle resource and the main component of ecology is the forest base that has been the main target of conservation. In addition, the success stories of forestry management such as community forestry are reported widely. Even though a good vegetation cover does retard the soil loss, geomorphic hazards related to flood and sediment are persistent in the region. In this regard, an issue raised by this study seems very important: the geomorphic processes are largely governed by many natural factors. Continuous expansion of the active gullies and changes in stream course in the floodplain are some of the examples to this. Even though human activities can be considered responsible for the changes in land cover, no clear relations could be identified between the human activities (mainly the deforestation) and geomorphic processes. It is likely that the changes in land cover may have impact on various erosion processes as dealt in this study, which could be overwhelmingly outweighed by the inherent geology and topography of the area. Considering this assertion, the management policy must be directed towards improving the land cover, at the same time introducing some sort of hard measures to lessen the impacts of soil erosion and overland flow.

What types of concrete policy/program are appropriate for dealing with the problems?

The Siwalik Hill ecology must be considered as the combination of upstream hill environment dominated by vegetation cover and the downstream floodplain mostly dominated by human activities. In addition to the requirements of maintaining a good ground cover in terms of forests and shrubs, engineering works are also necessary both in the upstream and downstream reach of the catchment as envisioned within the DPTC/JICA project. In particular, there is an urgent need for the stabilisation of active instabilities like landslides and gullies in the hills and stabilisation of stream banks in the terrace and flood plains.

The most important issue for such intervention schemes is the assessment of needs and available resources. Neither all the catchments in a region can be considered at the same time for the intervention, nor can it be considered all the stream channels within a catchment. For this, a systematic regional assessment approach would be necessary with the objective of identifying the most needy catchments and needy stream channels in the priority basis. In this regard, the results from this study are useful in two ways:

1. Status of catchment degradation can be evaluated based on the erosion rate and process map like presented in this study. Estimation of sediment yield from the catchment can be done taking the average rate values for each type of sediment source. Thus, the density of the sediment sources, mainly gullies and landslides (major sources) along with other sources such as surface and streambank erosion could be the basis for comparing the catchments within the area with relatively uniform geo-environments.

2. Once the catchment is identified and selected for detailed assessment, the technique such as stream bank erosion hazard mapping serves as an important tool for identifying the most vulnerable stream segments to lay priority for the protection works.

Being relatively small in spatial scale, the catchments of Siwalik Hills offer the best opportunity for the application of catchment-based integrated watershed management programme. The programme of works must focus on the schemes which can provide a quick and tangible benefit to the local people. For this following types of activities can be proposed:

- Strengthening stream banks by the combination of earthen bunds and vegetation, like one introduced by ChFDP.
- Riparian vegetation development using fodder trees and economic plants like bamboos.
- Gully control works by combining check dams and vegetation (such as bamboo) similar to a work undertaken in the Pipaltar area of Nuwakot district in the Middle Hills (Higaki et al., 1998).
- Additional efforts must be put on the stabilisation of active landslides and stream banks. Protection of talus slopes and promotion of vegetation can be a way forward for protection and conservation.

Summary

In summary, following points can be noted:

- Despite the recognition of the Siwalik Hills within the national plans as a sensitive region with constant threat of ecological degradation, there is much more to do on the ground especially to gain understanding of what works and what doesn't.
- Some conservation efforts such as by ChFDP and DPTC/JIA can be taken as good initiatives for the conservation of Siwalik Hills.
- High priority is given for forest management in the region, but field-based research works concerning gromorphological processes are lacking.
- Ecological degradation (in terms of water and sediment mobilisation) is mainly governed by natural factors such as geological composition. However, human intervention in terms of land use/ management is also necessary to lessen the effects from erosion activities.
- In addition to maintaining a good vegetation cover, efforts must be put for stabilising active instabilities such as gullies, landslides and riverbanks.
- Catchment-based integrated management of soil and water would be the best way forward for the management of Siwalik catchments in a sustainable way.

12
Lessons learned

Synthesis of the findings
Key results derived from this work are summarised in Table 12.1. Each study component has been described in terms of objectives, results and reasons or governing factors. Implications derived from the results are also summarised.

Lessons learned
In conclusion, this book attempts to address the following key research questions:

What are the main issues related to the environmental degradation in the Siwalik Hills?

The main issues identified included: floods, sediment mobilization, water scarcity and resource (mainly forest) degradation. Of them, floods and sediment related issues are incorporated in this book.

Stream floods are characterised by transient, sharp peaked, short duration, and rapid flow, which bring about miseries to the people directly in addition to many types of geomorphic hazards such as bank erosion and lateral shifting by inundation. The most concerned characteristic is that the stream channels are in the phase of development and are constantly widening. Particularly, the channel widening is evident around the floodplains.

Sediment related issues refer to the catchment degradation by various erosion processes, which is the main content of this study. Even though effects of sediment issues are less conspicuous, on-going sediment loss through various processes is one of the major concerns for ecological balance in the Siwalik Hills. The issue of vegetation degradation is closely attached to the problems related to sediments.

→ *Floods and sediment mobilisation, associated with land cover depletion are the major concerns of environmental degradation.*

What are the main processes related to sediment mobilization and how much is the contribution of each process to total sediment yield?

The main processes are classified as hillslope erosion and fluvial erosion. The first category consists of erosion from the surface including rills, gullies and landslide slopes and the second is mainly concerned with streambank erosion. There has been found a close relationship between the distribution of the sediment sources in the catchment. Landslides and gullies dominate in the hills whereas the active streambanks mostly occur in the terraces. As the hillslope erosion processes are very active, the bed of the stream channel is mostly filled with hillslope-derived sediments. The stream channels mostly act as transporting means, so the sediment from the bed erosion is likely to be insignificant compared to that from streambank erosion. This study is mainly concerned with the processes of sediment generation mostly in the headwater reach.

Table 12.1 Summary of key findings.

Study component	Objective	Results	Reasons/ governing factors	Implications
Land cover changes	Trend in the last four decades using sequential aerial/ satellite images (1964-2003)	-Trijuga river valley (649 km^2): (1964-2003)– Deforestation 17%. - Wide spatial and temporal variation -Khajuri and surrounding area (26.5 km^2) (1964-1992)- Deforestation 29% (1992-2003)- More or less no change	-Deforestation still persistent, especially in the inner (lower) Siwalik Hills (more favourable for agriculture the soil being finer in texture). -In the recent times, limited expansion of agriculture on the hillslope of Middle and Upper Siwaliks. -Improved community forestry management	- Land cover is only the component that can be changed from human activities. - Maintenance of vegetation cover in the headwater reach is important for lessening environmental degradation (especially surface erosion).
Stream pattern changes	Trend in the last four decades using sequential aerial/satellite images (1964-2003)	-Changing stream boundary mostly in the floodplain and confluence area -Trend: mostly widening (up to by 300% in floodplain) -Change in width: few meters to as high as 340m in the confluence - No direct linkage was detected between the change	- The streams are still in the phase of development. - In-stream process factors such as flow hydraulics, bed and bank materials etc. seem predominant over external factors. - Principle governing factors still not known.	- The time sequential overlay of stream boundary indicates the activity level of channel adjustment processes. Hence, it is a basis for identifying most unstable stream reach that need to be given priority for protection works.

Study component	Objective	Results	Reasons/ governing factors	Implications
		pattern of land cover and stream course.		
Rainfall-runoff relationship	Mechanism of flash flood	-The floods are ephemeral and transient types. - Time to peak: generally within 30 to 40 minute from the onset of flood -Duration: generally up to 5 hours, depends on duration of rainfall. - Runoff coefficient generally varies from 0 to 0.72.	- Dense and steep channel network in the catchment - Low infiltration rate (4-9 mm/h). - Steep slope gradient of the streams	- Highly erosive to streambanks due to high velocity. Some sort of velocity retarding structures should be introduced.
Types of sources	Relationship with geomorphological classification	-Relationship is close. Landslide and gully predominant in the hills while cutbanks in the terraces.	-Steep hills are composed of unconsolidated materials. - Fluvial deposits on the terraces	The classification system and distribution pattern can be applied in other areas in the Siwalik Hills.
Hillslope erosion processes	Identification of sediment sources and their rate and processes	Four processes dominate: surface, rill, gully and landslide		
Surface erosion	Average annual rate and operating processes	-Average annual rate: 1 ± 0.6 mm for dense forest, 0.9 ± 1.8 mm for partially dense forest,	-Effect of ground vegetation is more conspicuous than the effect of slope	Soil loss from bare soil is about 7 times higher than from the forest/shrub land. It

Study component	Objective	Results	Reasons/ governing factors	Implications
		1.2 ± 0.7 mm for shrubs and bushes, and 7 ± 3.8 mm for bare land. -Wide temporal and spatial variation in the erosion rates	gradient. - Low height plants and litter cover are more important factors.	indicates the need to preserve the vegetation cover in the hillslopes. Control of grazing is particularly important to maintain cover of low-height plants.
Rill erosion	Average annual rate and operating processes	-Vary widely in space in terms of shape and size. -Average rate of rill erosion on the bare slope (14 ± 6.8 mm/y) was found to be two times larger than the rate of surface erosion on the same slope. -Erosion-deposition phases in the rills formed in gentle slopes of shrubs and forest	-Closely related to runoff contributing area. - Fallen leaves and litters have significant influence in controlling erosion.	- Preservation of the fallen leaves and litters is important. For this, again importance of low-height plants is emphasized.
Gully erosion	Average annual rate and operating processes	From aerial photographs: -Head enlargement (1964 and 1992) by 34 to 58%. -Maximum retreat rates: 48 to 73 cm/year.	- Erosion amount not directly proportional to the rainfall amount. -Erosion largely depends on the stage of development. - Complexities	-Even though small in area, gullies are the important sources of sediment. -Gully protection works must be introduced in the head area

Study component	Objective	Results	Reasons/ governing factors	Implications
		- Estimated eroded volumes: 959 to 2783 m³/year. From field measurement: -Retreat rates: 4 to 28 cm/year. Eroded volumes:731± 57 to 2793 ± 201 m³/year. - Annual average sediment rates: 1164 ± 91 to 4100 ± 113 t/ha. - Process: dominance of mass failure from the tall headwalls and channel sidewalls.	in headwall and drainage channel erosion pattern especially in multi-bed geological formation	than in the main channel.
Landslide slope erosion	Average annual rate and operating processes	-The steep hillslopes exhibit a large density of landslides, both active and stabilized. -Rill erosion is p the slide slope s -The erosion varies from 5 ± 1 to 6 ± 3 cm/year. Principle of parallel retreat is assumed in estimating sediment	-Mainly induced by stream channel undercutting - Type, shape and size obviously vary in relation to the geology.	-Landslides in terms of gradual soil loss are very complex phenomenon. - Considering the parallel retreat of slope erosion, monitoring of surface erosion, like done in this study is one way to estimate sediment production approximately.

Study component	Objective	Results	Reasons/ governing factors	Implications
Fluvial erosion process: Streambank erosion	Average annual rate and operating processes	- About 40% of the bank is active with cutslope. -Erosion rates: few cm to as high as 2 m/year. -In the forested area, average erosion :12 cm/year. -Estimated sediment volume: about 524 ± 245 t / km^2/year (in catchment scale). -Process: cantilevered or overhanging banks are generated when erosion of an erodible layer leads to undermining of overlaying, erosion-resistant layer. - Evaluation of bank hazard is done by using the rating of bank variables.	- Isolated effects of controlling factors are difficult to determine. However predominant factor seems the bank material composition. - Vegetation can have both positive and negative effects on bank stability: densely vegetation generally strengthens the stream banks however isolated trees may weaken the banks with the action of turbulence and local scouring.	- Maintaining a dense bank vegetation is important for reducing bank erosion. -The methodology, which has been tested to evaluate the bank erosion hazard, can be applied to other streams in the Siwalik Hills with proper modification. - Its combination with the stream planform overlay can be a good basis for evaluating and verifying the hazard level.
Relative contribution of sediment sources	Assessment of relative contribution of each sediment sources in the catchment scale.	-Prime erosion rates in catchment: Surface erosion: Forest/shrub-	- The erosion figures are the average ones over the monitoring period, which are	- Contribution of landslides is the greatest because of high densities of landslides. - In terms of

Study component	Objective	Results	Reasons/ governing factors	Implications
		1mm (14 t/ha) Agriculture (estimated)- 2mm (28 t/ha) Rill erosion: not accounted for. Gully erosion: 3335 t/ha (of gully area) Landslide erosion: 1120 t/ha (of landslide area) Bank erosion: 367 t/km - Relative contribution (in catchment): Surface erosion- 22% Gully erosion- 18% Landslide erosion- 48 % Bank erosion- 12 %	extrapolated in the catchment scale. - Rill erosion is not accounted for because of the lack of information on the rill density in the catchment. - Channel bed erosion is also not taken, because field observation revealed that majority of the sediment is produced from bank erosion.	areal context, gullies are the important sediment sources which have greatest sediment yield rates per unit area. Hence, priority must be put on checking the sediment from gullies, as well as landslides. - Bank erosion has least contribution, but can have important local effects.
Sediment estimation	Estimation of total sediment production from the catchment.	-Catchment area: 4.62 km^2 -Average production rate: 7458 t/km^2 (equivalent to the denudation rate of 5.3 mm) -Measured suspended	-Sediment delivery ratio for small and steep catchment tends to be unity. In addition, as the gravel and boulders are of fluvial origin, erodibility of the bed material may be higher because of low	- More precise sediment budget can be estimated taking into account of bedload. - The sediment production rates from each type of sources and the catchment as whole, and the distribution map of

Study component	Objective	Results	Reasons/ governing factors	Implications
		sediment rate: 6750-7350 t/km². -Catchment-scale distribution map of erosion rates and processes has been prepared.	frictional resistance.	erosion rates and processes can be an important basis for the designing of conservation works.
Evaluation of catchment status	To evaluate the degradation status of the catchment.	- When compared to the results from other studies, the catchment can be regarded as severely degraded.	- The erosion rates are well within the ranges suggested by many studies in the Siwalik Hills.	- These all observations and analysis indicate that the catchments of the Siwalik Hills are truly in the degraded status. - However, the erosion processes are essentially natural. Human intervention is necessary to lessen the activities of the processes.

Each type of process has its own significance in terms of causes and effects in the geomorphic environment. However, in terms of sediment yield from the catchment, landslides are found to be the major sources of sediment, which generate about half of the total sediment from the study catchment. Gullies are also the important point sources of sediment in the sense that the sediment yield per unit area is highest. In the study catchment, their aggregate contribution to the total sediment is found to be about 18%. Surface and streambank erosion accounts for about 22 and 12% respectively.

→ *Hillslope erosion (surface with rill, gully and landslide slope) and fluvial erosion (mostly streambank erosion) are the major processes of sediment generation.*

→ *Landslide slopes and gullies are the major point sources of sediment.*

What are the governing factors of these erosion processes and what is the effect of human activities?

The hillslope and fluvial erosion processes are the results of combined rather than individual factors. Isolation of the individual effect of a particular governing parameter is almost not possible because of high degree of complexity in the process actions.

The governing factors and relevant processes are shown in Fig. 12.1. Rainfall, which is the main input to the catchment system is intense and bigger compared to other geophysical environments. Geology is particularly weak and unconsolidated as a result of relatively new formation. The original bedrocks are of fluvial origin, which are developed by crustal movement. Topography is characterized by extremely steep hillslopes and thus the drainage channels are steep too. The soil matrix is weak and shallow with low infiltration making the slopes vulnerable to many types of instabilities. All these hostile factors combine to make the hillslope and fluvial erosion processes more active. Surface erosion is moderate compared to the other types of sediment processes mainly because of good vegetation cover. However, the human activities are exerting constant pressure on the forest resource.

In the case of Siwalik Hills, the inherent natural characteristics described above seem to have more effects on the erosion processes compared to the human activities. While the effects of human activities are clearly evident in terms of land cover change; further effects in terms of geomorphologic changes downstream could not be identified clearly.

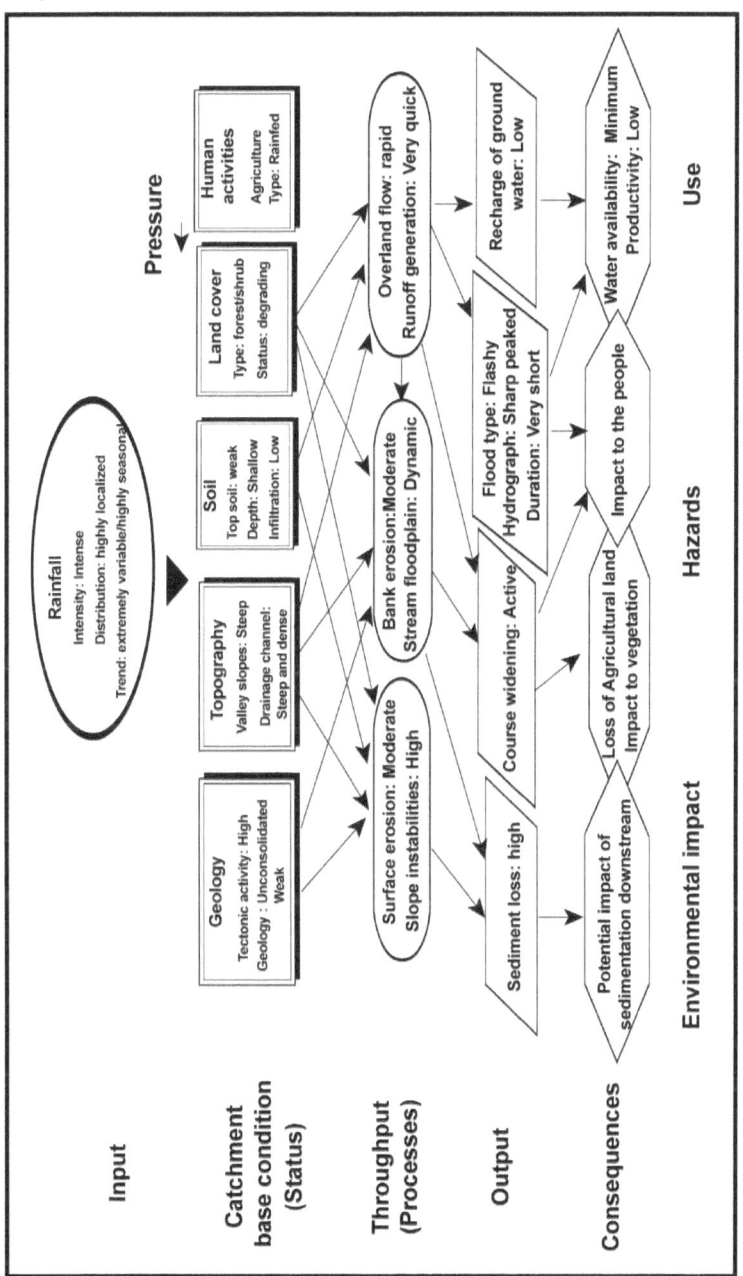

Figure 12.1 Correlation matrix showing the key catchment processes, driving factors and consequences in the Siwalik Hills.

→ *While there is constant pressure of human activities especially on land cover change (deforestation), the natural factors tend to overweigh the effects in terms of sediment generation.*

What is the degradation status of the catchment?

From the viewpoint of land cover, majority of the catchment is covered with forest/shrubs, which may indicate a better status of the catchment. However, from the standpoint of sediment generation, the case is not the same. From the study, it is found that average rate of catchment sediment production rate is about 7500 $t/km^2/year$ which is equivalent to the denudation rate of 5 mm per year. Compared to the other areas in Nepal, the denudation rate is very high. It indicates that the catchment is in the state of severe degradation. It is mainly because of the high density of landslides within the catchment. In addition, the occurrence of active gullies makes the condition worst.

→ *Based on the sediment production, the catchment can be viewed as severely degraded.*

What types of management policies are appropriate for addressing the soil and water conservation issues in the Siwalik Hills?

In spite of the less contribution of surface erosion towards the total sediment yield, it is still vital to conserve and promote the vegetation cover on the steep hillslope in the upstream reach. The management policies must be directed towards the catchment-based integrated approach where some sort of structural countermeasures might be necessary to stabilise active instabilities like gullies and landslides. At the same time, structural countermeasures must be introduced in the downstream reach to help strengthen the stream banks and confine the stream channels. Combination of vegetation with the structural works does not only increase the effectiveness but also provides multiple benefit to the local people. Field-based action research works must be promoted to understand the process interactions within the catchment system.

→ *The ecological degradation in the Siwalik Hills should include two components simultaneously: upstream land degradation and downstream geomorphic hazards.*

→ *Micro-scale watershed management based on integrated approach should be included in the conservation policy.*

Finally, this study has documented the ecological importance of the Siwalik Hill region in terms of upstream-downstream linkage with the Terai plain which is the back bone of agriculture-based economy of the country. Hence, preservation of the Siwalik Hills means preservation of the Terai plain too.

→ *The Siwalik Hill region is the most important ecological environment that has broader significance in terms of cause and effect at an inter-regional scale.*

Recommendations

Some of the themes that can be considered in the future:

- Investigation on erosion processes in the event basis. From this, effects of rainfall size, intensity and duration on the erosion rates and processes can be understood.
- Detailed process study on landslide slope erosion. From this, erosion activities and sediment production from the landslides can be identified.
- Further modification of the bank erosion hazard evaluation technique developed in this study. The technique should be tested in other streams too in order to evaluate its applicability and limitation.
- Evaluation of the effects of land cover changes on the geomorphological changes in other areas of region. Land cover changes and its effects on the hydro geomorphological environments are the current issues still unsolved.
- Evaluation of the structural and vegetation techniques for controlling the erosion. This must be carried out with the help of some field-based action research works. This can be helpful in testing the applicability of conservation works in a specific socio-economic and geo-physical environment.
- Comparative study on the geomorphological development of catchments situated on the south and north face of the Siwalik Hills. Since the geology consists mostly of north-dipping rocks, there must be some differences such as development patterns of drainage channel and associated erosion patterns.

- Study on upstream-downstream linkages between the Siwalik Hills and Terai plain. This study can help to understand the causes and effects relationship mainly in the water and sediment mobilization. Since the Siwalik Hills are the major sources of sediment that the stream systems carry towards the plain of Terai, such type of inter-regional study will be of much importance.

References

Abrahams, A.D.; Li, D.G. and Parsons, A.J.(1996). Rill hydraulics on a semi-arid hillslope, southern Arizona. Earth Surface Processes and Landforms (21):35-47.

Ashbridge, D., (1995). Processes of river bank erosion and their contribution to the suspended sediment load of the River Culm, Devon. In Sediment and Water Quality in River Catchment Systems, Foster IDL, Gunell AM, Webb BW (eds.). Wiley: Chichester: 229-245.

Bull, L. (1997). Magnitude and variation in the contribution of bank erosion to the suspended sediment load of the river Severn, UK. Earth Surface Processes and Landforms (22) :1109-1123.

Burkard, M.B. and Kostaschuk, R.A. (1997). Patterns and controls of gully growth along the shoreline of lake Huron. Earth Surface Processes and Landforms (22): 901-911.

CBS (2001). Statistical year book of Nepal 2001.Central Bureau of Statistics, Kathamndu.

Chalise,S.R.; Shrestha, M.L.; Thapa, K.B.; Shrestha, B.R. and Bajracharya, B. (1996). Climatic and hydrological atlas of Nepal, Kathmandu: ICIMOD.

Chaudhary, R.P. (1998). Biodiversity in Nepal: Status and conservation. S.Devi, Saharanpur, India and Tecpress Books, Bangkok.

ChFDP (2001). Churia Forest Development Project (Brochure 2001), Siraha, Lahan.

Chow, V.T. (1951). "A general formula for hydrologic frequency analysis. Trans. Am. Geophysical Union (32) no.2: 231-237.

Chow, V.T., Maidment, D.R., Mays,L.W.(1988). Applied Hydrology. McGraw-Hill Book Company, NY.

Couper, P., Stott, T. and Maddock, A., (2002). Insights into river bank erosion processes derived from analysis of negative erosion-pin recordings: observations from three recent UK studies. Earth Surface Processes and Landforms (27): 59-79.

Couper, P.R. and Maddock, I.P.(2001). Subaerial river bank erosion processes and their interaction with other bank erosion mechanisms on the river Arrow, Warwickshire, UK. Earth Surface Processes and Landforms (26): 631-646.

Darby, S.E. and Thorne, C.R. (1996). Development and testing of riverbank stability analysis. Journal of Hydraulic Engineering(122), 8.: 443-454.

Derose, R.C.; Gomez, B.; Marden, M. and Trustrum, N.A. (1998). Gully erosion in Mangatu forest, New Zealand, estimated from digital elevation models. Earth Surface Processes and Landforms (23): 1045-1053.

Dhakal and Sidle, (2004). Distributed simulations of landslides for different rainfall conditions. Hydrological Processes (18): 757-776.

DHM (2003). Rainfall data records of Udayapur in 2002 and 2003. Department of Hydrology and Meteorology, Kathmandu.

DOF (1995). Community Forestry Directives 1995 Kathmandu: HMG/MOFSC Department of Forests.

DSCWM (2004). Watershed Management Policy and Strategies. Department of Soil Conservation and Watershed Management. http://www.nepalhmg.gov.np/dscwm/policy.html

Ebisemiju, F.S. and Ekiti, A. (1989). A morphometric approach to gully analysis. Z. Geomorph. N.F. (33-3): 307-322.

Eckholm, E. (1976). Losing ground. Environmental stress and world food prospects. New York: W.W. Norton & Company.

FAO (2002). Land-water linkages in rural catchments. FAO Land and Water Bulletin 9, Food and Agriculture Organization, Rome.

Gardiner, T.,(1983). Some factors promoting channel bank erosion, River Lagan, Country Down. Journal of Earth Science Royal Dublin Society(5): 231-239.

Gautam, A.P., Webb, E.L. and Elumnoh, A. (2002). Assessment of land use/land cover changes associate with community forestry implementation in the Middle Hills of Nepal. Mountain Research and Development, 22(1): 63-69.

Ghimire, S.K, Higaki, D. and Bhattarai, T.P. (2003). Flash floods in the Siwaliks: Consequences and countermeasures, A case study in Nepal. Proc.of 1st International conference on Hydrology and Water Resources in Asia Pacific Region (APHW), 13-15 March, Kyoto Japan: 833-838.

Ghimire, S.K. and Higaki, D. (2004). Changes in landuse and stream planform: Implications for the sustainable management of ephemeral streams in Siwalik Hills, Nepal. Proc. of the 2nd Asia Pacific Association of Hydrology and Water Resources (APHW) conference, 5-8 July, Singapore: 261-270.

Ghosh, R.C. and Subba Rao, B.K.(1979). Forests and floods. Indian Forester (105): 249-259.

Gilvear, D., Winterbottom, S. and Sichingabula, H.(2000). Character of channel planform change and meander development: Luangwa river, Zambia. Earth Surface Processes and Landforms (25): 421-436.

Govers,G. (1985). Selectivity and transport capacity of thin flows in relation to rill erosion. Catena (12): 517-524.

Graf., W.L. (1984). A probabilistic approach to the spatial assessment of river channel instability. Water Resour. Res. (20)7: 953-962.

Green, T.R., Beavis, S.G., Diertich, C.R.and Jakeman, A.J. (1999). Relating stream bank erosion to in-stream transport of suspended sediment. Hydrological Processes (13): 777-787.

Hasegawa, K.(1989). Universal bank erosion coefficient for meandering rivers. Journal of Hydraulic Engineering(115) 6.: 744-765.

Heed, B.H. (1975). Watershed indicators of land-form development,. In Hydrology and water resources in Arizona and the Southwest (5), American Water Resources Association and Hydrology Section, Arizona Academy of Science: 43-46.

Heede, H. (1975). Stages of development of gullies in the west. In Present and Prospective Technology for Predicting Sediment Yields and Sources, Proceedings of the Sediment-Yield Workshop, USDA, Oxford, Mississippi.

Hickin, E.J. and Nanson, G.C.(1984). Lateral migration rates of river bends. Journal of Hydraulic Engineering (110):1557-67.

Higaki, D. (1998). Erosion and sedimentation problems in Nepal from the viewpoint of morphological development. Journal of Nepal Geological Society (18): 309-318.

Higaki, D. (2003). Landslides and erosion study in Siwalik region using geomorphological approach. 1st Seminar of Nepal Landslide Society (NLS), Kathmandu.

Higaki, D.; Karki, K.K. and Gautam, C.S.(1998). Soil erosion problems in the midland of Nepal with reference to the Trisuli model site, Nuwakot district, Central Nepal. News Bulletin of Nepal Geological Society (15): 33-38.

HMG (2002). Policies and Programmes for Poverty Alleviation (10th plan). His Majesty`s Government of Nepal.

HMG/ADB/FINNIDA(1988). Master Plan for the Forestry Sector: Main Report. Kathmandu.

Honda, K., Samarakoon, L.,Ishibahi, A., Mabuchi, Y. and Miyajima, S.(1996). Remote sensing and GIS technologies for denudation estimation in a Siwalik watershed of Nepal. Proc. Intl Seminar on Water Induced Disaster (ISWID), Katmandu: 294-301.

Horton, R.E. (1939). Analysis of runoff plat experiments with varying infiltration capacity. Trans. Am. Geophys. Union, (20): 693-711.

Houghton, R.A. (1994). The worldwide extent of land use change. Bio Science 44 (5): 305-313.

Huang, H.Q. and Nanson, G.C. (1998). The influence of bank strength on channel geometry: An integrated analysis of some observations. Earth Surface Processes and Landforms (23): 865-876.

ICIMOD (1996). Climatic and Hydrological Atlas of Nepal. International Centre for Integrated Mountain Development).Katmandu, Nepal.

Irasawa, M. (2004). Soil survey report of Khajuri stream (unpublished). Iwate University, Japan.

Itihara, M., Shibazaki, T. and Miyamoto, N. (1972). Photogeological survey of the Siwalik Ranges and the Terai Plain, southeastern Nepal. Journal of Geoscience (15), Osaka City University: 77-99.

Ives, D.J. and Messerli, B.(1989). The Himalayan Dilemma-Reconciling Development and Conservation, London, Routledge.

Kimura, K. (1997). Morphotectonic development of the western part of the Trijuga dun, Nepal Sub-Himalaya. Science Report of Tohoku University, vol. 47, no.1/2, Institute of Geography, Tohoku University, Japan.

Kimura, K. (2000): Morphostructural sequence and active tectonics of the Himalayan Front, Journal of Geography, 109(1), p: 56-72.

Kirkby, M.J. and Morgan, R.P.C. (1980). Soil Erosion, Wiley-Interscience, London and New York.

Kizaki, K. (1988). Rising Himalayas (in Japanese). Tsukiji Shokan, Tokyo.

Knighton, A. David, Nanson, Gerald C. (2001). An event-based approach to the hydrology of arid zone rivers in the Channel Country of Australia. Journal of Hydrology (254):102-123.

Knighton, D. (1998). Fluvial forms and processes, a new perspective. Arnold, London NW13BH.

Koning, G.H.J, Verburg, P.H., Veldkamp, A., Fresco, L.O. (1999). Multi-scale modelling of land use change dynamics in Ecuador. Agricultural Systems 61: 77-93.

Kukal, S.S., Sur, H.S. and Gill, S.S.(1991). Factors responsible for soil erosion hazard in submontane Punjab, India. Soil Use & Management (1) : 38-44.

Laban, P.(1978). Field measurements of erosion and sedimentation in Nepal. Dept. of Soil Conserv. and Watershed Man., Katmandu, Paper no. 5.

Lauterburg, A.(1993). The Himalayan highland-lowland interactive system: Do landuse changes in the Mountains affect the plains?. In, Messerli, B; Hofer, T.; Wymann, S.(eds.). Himalayan Environment Pressure-Problems-Processes: 21-30. Institute of Geography, University of Bern.

Lawler D. M. (1993). The measurement of river bank erosion and lateral channel change: a review. Earth Surface Processes and Landforms. Technical Software Bulletin, (18): 777–821.

Lawler, D.M., Couperthwaite, J., Bull, L. and Harris, N.M. (1997). Bank erosion events and processes in the Upper Severn Basin. Hydrology and Earch Surface Sciences(1):523-534.

Lawler, D.M., Grove, G.R., Couperthwaite, J.S. and Leeks, G.L. (1999). Downstream change in river bank erosion rates in the Swale-Ouse system, northen England. Hydrological Processes (13): 977-992.

Lawler, D.M., Thorne, C.R. and Hooke, J.M. (1997). Bank erosion and stability. In Applied Fluvial Geomorphology for River Engineering and Management, Thorne CR, Hey RD, Newson MD(eds.). Wiley:Chichester: 137-172

Leopold, L.B, Wolman, M.G. and Miller, J.P. (1964). Fluvial processes in geomorphology. W. H. Freeman and company, San Francisco.

LRMP (1986). Land Systems, Land Utilization and Agriculture-Forestry Reports. Land Resource Mapping Project. Ottawa, Canada: Kenting Earth Sciences Ltd.

Martfnez- Casasnovas, J.A.; Anton-Fernandez, C. and Ramos, M.C. (2003). Sediment production in large gullies of the Mediterranean area (Ne Spain) from high-resolution digital elevation models and geographical information systems analysis. Earth Surface Processes and Landforms (28): 443-456.

Martin-Penela, A.J.(1994). Pipe and gully systems development in the Almanzora basin (Southest Spain). Z. Geomorph. N.F. (38-2): 207-222.

Merz, Y. (2004). Water balances, floods and sediment transport in the Hindu Kush-Himalayas: Data analyses, modeling and comparison of selected meso-scale catchments. P.Hd. Thesis. Institute of Geography, University of Berne, Switzerland.

MFSC (2000). Forest Sector Policy 2000. Ministify of Forest and Soil Conservation.

MFSC(1999). Forest and Shrub Cover of Nepal (1989-1994). Ministry of Forests and Soil Conservation, Department of Forest Survey and Research, Kathmandu.

Micheli, E.R. and Kirchner, J.W. (2002). Effects of wet meadow riparian vegetation on streambank erosion. 2. Measurements of vegetated bank strength and consequences for failure mechanics. Earth Surface Processes and Landforms (27)7: 687-697.

Micheli, E.R. and Kirchner, J.W. (2002). Effects of wet meadow riparian vegetation on streambank erosion. 1. Remote sensing measurements of streambank migration and erodibility. Earth Surface Processes and Landforms (27),6: 627-639.

Mittal, S.P., Bansal, R.C., Sud, A.D. and Dyal, S.N. (1984). Hydrological behaviour of agricultural watershed. Annual report, Central Soil and Water Conservation Research and Training institute, Dehradun, India:83-84.

MOFSC (2002). Ministry of Forest and Soil Conservation http://www.geocities.com/watershed_nepal/

MOFSC (2002). Nepal Biodiversity Strategy – 2002. Ministry of Forest and Soil Conservation.

MOFSC (1994). Deforestation in the Terai districts, Forest resource information system project, Ministry of Forest and Soil Conservation.

MOPE (2002). World Summit on Sustainable Development (Rio+10), National Assessment Report 2002. Ministry of Population and Environment,. http://www.mope.gov.np/environment/assessment.php

MOPE(2000). Nepal's State of the Environment. His Majesty's Government, Ministry of Population and Environment, Kathmandu.

Morgan, R.P.C., (1986). Soil Erosion and Conservation. Addison Wesley Longman. Harlow, England.

Nachtergaele, J. and Poesen, J. (1999). Assessment of soil losses by ephemeral gully erosion using high-altitude (Stereo) aerial photographs. Earth Surface Processes and Landforms (24): 693-706.

Nachtergaele, J.; Poesen J; Vandekerckhove, L.; Oostwoud Wijdenes, D.J.; and Roxo, M. (2001). Testing the ephemeral gully erosion model (EGEM) for two Mediterranean environments. Earth Surface Processes and Landforms (26):17-30.

Nakata, T.(1982). A photographic study on active Faults in the Nepal Himalayas. Journal of Nepal Geological Society (2): 67-80.

Nearing, M.A.; Norton, L.D.; Bulgakov, D.A.; Larionov, G.A.; West, L.T.; Dontsova K.M.(1998). Hydraulics and erosion in eroding rills. WaterResources Research (33): 865-876.

Oli, K.P.(1999). Siwalik - the Threatened Natural Heritage. IUCN-Nepal Newsletter 3 (3):1-6

Oliveira, M.A.T. (1990). Slope geometry and gully erosion development: Bananal, Sao Paulo, Brazil. Z. Geomorph. N.F. (34-4): 423-434.

Oostwoud Wijdenes, D.J. and Bryan, R.(2001). Gully-head Erosion Processes on a semi-arid valley floor in Kenya: A case study into temporal variation and sediment budgeting. Earth Surface Processes and Landforms (26): 911-933.

Oostwoud Wijdenes, D.J.; Poesen, J. ; Vandekerckhove, L.; Nachtergaele, J. and Baerdemaeker, J. (1999). Gully head morphology and implications for gully development on abandoned fields in a semi-arid environment, Sierra de Gata, Southeast Spain. Earth Surface Processes and Landforms (24):585-603.

Patton, P.C. and Schumm, S.A. (1981). Ephemeral stream processes: implications for studies of Quaternary valley fills. Quaternary Research (15): 24-43.

Pfaff, A.S. (1999). What drives deforestation in the Brazilian Amazon? Journal of Environmental Economics and Management, 37: 26-43.

Piest, R. F., Bradford, J. M. and Spomer, R. G.(1975). Mechanisms of erosion and sediment movement from gullies. In Present and Prospective Technology for Predicting Sediment Yields and Sources, Proceedings of the Sediment-Yield Workshop, USDA, Oxford, Mississippi, 1972, 162–176.

Poesen J, Nachtergaele J, Verstraeten G and Valentin C. (2003). Gully erosion and environmental change: importance and research needs. Catena (50): 91-133.

Rosgen, D.L.(2001). A practical method of computing streambank erosion rate. Wildland Hydrology. Inc. Colorado. http:/www.wildlandhydrology.com

Samarakoon, L.,Ishibahi, A., Mabuchi, Y., Honda, K., and Miyajima, S.(1996): Investigating river channel changes as a consequence of continuous flood hazard in Terai region of Nepal using remotely sensed data. Proc. Intl Seminar on Water Induced Disaster (ISWID), Katmandu, p: 294-301.

Schelling, D. (1992). The tectonostratigraphy and structure of the Eastern Nepal Himalaya. Tectonics (11): 925-943.

Schumm, S.A. (1973). Geomorphic thresholds and complex response of drainage systems. In Morisawa, M. (ed.), Fluvial Geomorphology. Binghamton, New York: 299-309.

Schumm, S.A.; Harvey, M.D. and Watson, C.C. (1984). Incised Channels. Water Resources Publications, Colorado: 200-208.

Schweik, C.M., Adhikari, K. and Pandit, K.N. (1997). Land-cover changes and forest institutions: a comparison of two sub-basins in the Southern Siwalik Hills of Nepal. Mountain Research and Development, 17:99-116.

Seginer, I.(1966). Gully development and sediment yield. Journal of Hydrology (4): 236–253.

Selby, M.J.(1993). Hillslope Materials and Processes. Oxford University Press, Oxford.

Shah, P.B., Schreier, H., Nakarmi, G. (2000). Rehabilitation of degraded lands. In, Allen, R., Schreier, Brown, S., Shah, P.B. (eds.). The People and Resource Dynamic Project. The first three years (1996-1999). Proceedings of a workshop , Baoshan, 2-5 March 1999: 139-148

Sharma, C.K. (1981). Landslides and soil erosion in Nepal.Ms. Sangeeta Sharma, Bishalnagar, kathmandu.

Sharma, C.K., (1990): Geology of Nepal Himalaya and Adjacent Countries. Sangeeta Sharma, Kathmandu, Nepal.

Sharma, C.K. (1977). River systems of Nepal. Kathmandu, Sangeeta Sharma.

Shrestha, B. and Brown, S. (1995). Land use dynamics and intensification. In Schreier, H., Shah, P.B. and Brown, S. (eds.). Challenges in mountain resource management in Nepal: Processes, Trends and Dynamics in Middle Mountain watershed. Workshop proceedings. International Centre for Integrated Mountain Development (ICIMOD) Katmandu, Nepal: 141-154.

Simon, A. and Collison, A.J.C.(2001). Pore-water pressure effects on the detachment of cohesive streambeds: seepage forces and matric suction. Earth Surface Processes and Landforms (26):1421-1442.

Simon, A. and Throne, C.R.(1996). Channel adjustment of an unstable coarse-grained alluvial stream: opposing trends of boundary and critical shear stress and the applicabilty of external hypothesis. Earth Surface Processes and Landforms (21).155-180.

Simon, A. (1989). The discharge of sediment in channelized alluvial streams. Water Resources Bulletin (25): 1177-1188.

Simon, A., Curini, A., Darby, S.E. and Langendoen, E.J. (2000). Bank and near-bank processes in an incised channel. Geomorphology (35): 193-217.

Simon, A., Wolfe, W.J. and Molinas, A.(1991). Mass wasting algorithms in an alluvial channel model. Proc. Of the fifth Federal Interagency Sedimentation conferece, Las Vegas, Nevada, vol. 2 : 71-82.

Simon. A. and Collison, A.J.C.(2002). Quantifying the mechanical and hydrologic effects of riparian vegetation on streambank stability. Earth Surface Processes and Landforms (27),5: 527-546 .

Singh, G., R. Babu, P. Narain, L. S. Bhushan, and I. P. Abrol. 1992. Soil erosion rates in India. Journal of Soil and Water Conservation 47 (1): 97-99.

Singh, Y. (2001). Geo-ecology of the Trans Satluj Punjab-Haryana Siwalik Hills, NW India. Himalayan Ecology & Development, (9), No. 2.

Skole, D.L., Chomentowski, W.H., Salas, W.A., Nobre, A.D., (1994). Physical and human dimensions of deforestation in Arnazonia. BioScience 44 (5): 314-322.

Slattery, M.C. and Bryan, R.B.(1992). Hydraulic conditions for rill incision under simulated rainfall: a laboratory experiment. Earth Surface Processes and Landforms (17):127-146.

Smith, D.G. (1976). Effect of vegetation on lateral migratin of a glaier meltwater river. Geological Society of America Bulletin (87):857-860.

Stevens,S.F. (1993). Claiming the high ground: Sherpas, Subsistence and environmental change in the Highest Himalaya. University of California Press, Berkeley, USA.

Stocking, M. A.(1980). Examination of the factors controlling gully growth. In DeBoodt, M. and Gabriels, D. (Eds), Assessment of Erosion, John Wiley & Sons, Chichester, 505–520.

Stott, T.(1997). A comparison of stream bank erosion processes on forested and moorland streams in the Balquhidder catchments, Central Scotland. Earth Surface Processes and Landforms (22): 383-399.

Subba Rao,B.K., Dabral, B.G., and Ramola, B.C. (1973). Quality of water from forested watersheds. Indian Forester (99):681-690.

Tekle, R. and Redlund, L. (2000). Land cover changes between 1958 and 1986 in Kalu district, Southern Wello, Ethiopia. Mountain Research and Development, 20(1): 42-51.

Tervuren, J.M. (1990). Soil loss by rainwash: a case study from Rwanda. Z. Geomorphologie N.F., 34 (4): 385-408.

Thomas, A.W. and Welch, R. (1988). Measurement of ephemeral gully erosion. Transactions of the ASAE, 31(6):1723-1728.

Thomas, M.F. (2001). Landscape sensitivity in time and space- an introduction. Catena, 42: 83-98.

Thompson, J. R.(1964). Quantitative effect of watershed variables on rate of gully-head advancement, Transactions of the ASAE, 7(1): 54–55.

Thorne, C.R. (1997). Channel types and morphological classification. In Applied Fluvial Geomorphology for River Engineering and Management, (eds) Thorne, C.R., Hey, R.D., Newson, M.D., John Wiley and Sons, New York.

Thorne, C.R. (1990). Effects of vegetation on riverbank erosion and stability. In Vegetation and Erosion, Thornes J.B.(ed.). Wiley:Chichester:125-144.

Tole, L. (1998). Sources of deforestation in tropical developing countries. Environmental Management, 22: 19-33.

Upreti, B.N. and Le Port, P. (1999). Geology of the Nepal Himalaya: Recent advances. Journal of Asian Earth Sciences 17(5-6): 577-606.

Upreti, B.N.(2001). The physiographic and geology of Nepal and their bearing on the landslide problem. In, Landslide Hazard Mitigation in the Hindu Kush- Himalaya. Tianchi, L.; Chalise, S.R.; Upreti, B.N. (eds.). ICIMOD, Nepal.

USDA (2003). Bank erosion and channel widening. United States Department of Agriculture. Http://www.sedlab.olemiss.edu/agnps/concepts/manual/streambank.html

Valdiya, K.S. (1998). Dynamic Himalaya. Hyderabad: Universities Press.

Vandaele, K.; Poesen, J.; Govers, G. and Van Wesemael, B. (1996). Geomorphic threshold conditions of ephemeral gully incision. Geomorphology (16): 161-173.

Vandekerckhove L, Poesen J, Oostwoud Wijdenes DJ, Gyssels G. (2001). Short-term bank gully retreat rates in Mediterranean environments. Catena 44(2): 133-161.

Vandekerckhove, L.; Poesen, J. ; Oostwoud Wijdenes, D.J.; Nachtergaele, J.; Kosmas. C., Roxo, M.J. and Figueiredo, T. (2000). Thresholds of gully initation and sdimentation in Mediterranean Europe. Earth Surface Processes and Landforms (25):1201-1220.

Vandekerckhove, L.; Poesen, J. and Govers, G.(2003). Medium-term gully headcut retreat rates in Southeast Spain determined from aerial photographs and ground measurements. Catena (50): 329-352.

Vandekerckhove, L.; Poesen, J.; Oostwoud Wijdenes, D.J; Gyssels, G., Beuselinck, L. and Luna, E. (2000). Characteristics and controlling factors of bank gullies in two semi-arid Mediterranean environments. Geomorphology (33): 37-58.

Varnes, D.J. (1958). Landslide types and processes. Highway Research Board, Special Report (Washington, DC) 29: 20-47.

Walling, D.E. (1983). The sediment delivery problem. Journal of Hydrology (65): 209-237.

Wasson RJ, Mazari RK, Starr B, Clifton G. (1998). The recent history on erosion and sedimentation on the Southern Tablelands of southeastern Australia: sediment flux dominated by channel incision. Geomorphology (24): 291-308.

WECS (1987). Erosion and sedimentation in the Nepal Himalaya. Water and Energy Commission Secretariat. Ministry of Water Resources, Kathmandu.

Winterbottom, S. and Gilvear, D. (2000). A GIS-based approach to mapping probabilities of river bank erosion: Regulatedd river Tummel, Scotland. Regulated Rivers: Research and Management (16): 127-140.

Wood, A.L., Simon, A., Downs, P.W. and Throne, C.R. (2001). Bank-toe processes in incised channels: the role of apparent cohesion in the entrainment of failed bank materials. Hydrological Processes (15):39-61.

Xu, J. (2002). Complex behaviour of natural sediemnt-carrying streamflows and geomorphological implications. Earth Surface Processes and Landforms (27),7: 749-758

Young, A. (1972). Slopes. Longman Group Ltd. London and New York.

Zomer, R.J., Ustin, S.L. and Carpenter, C.C. (2001). Land cover change along tropical and subtropical riparian corridors within the Makalu Barun National Park and conservation area, Nepal. Mountain Research and Development, 21(2): 175-183.

www.ingramcontent.com/pod-product-compliance
Lightning Source LLC
Chambersburg PA
CBHW030942180526
45163CB00002B/679